HANDBOOK OF
ADHESIVE BONDED STRUCTURAL REPAIR

HANDBOOK OF ADHESIVE BONDED STRUCTURAL REPAIR

by

Raymond F. Wegman and Thomas R. Tullos

Adhesion Associates
Ledgewood, New Jersey

np

NOYES PUBLICATIONS
Park Ridge, New Jersey, U.S.A.

Library of Congress Catalog Card Number: 91-46660
ISBN: 0-8155-1293-7
Printed in the United States

Published in the United States of America by
Noyes Publications
Mill Road, Park Ridge, New Jersey 07656

10 9 8 7 6 5 4 3 2 1

Library of Congress Cataloging-in-Publication Data

Wegman, Raymond F.
Handbook of adhesive bonded structural repair / by Raymond F. Wegman and Thomas R. Tullos.
p. cm.
Includes bibliographical references and index.
ISBN 0-8155-1293-7
1. Sandwich construction. 2. Adhesive joints--Repairing. 3. Composite materials. 4. Laminated materials. I. Tullos, Thomas R. II. Title.
TA492.S25W45 1992
624.1'8--dc20 91-46660
CIP

PREFACE

The purpose of this book is to provide information and guidance on the repair of adhesive bonded and composite structures. The methods presented have been generally published in the open literature. However, these methods have not been verified by the authors for every application. It is the user's responsibility to evaluate any selected method for its adaptability to the structure to be repaired.

Raymond F. Wegman

NOTICE

To the best of our knowledge the information in this publication is accurate; however, the Publisher does not assume any responsibility or liability for the accuracy or completeness of, or consequences arising from, such information. Mention of trade names or commercial products does not constitute endorsement or recommendation for use by the Publisher.

Final determination of the suitability of any information or product for use contemplated by any user, and the manner of that use, is the sole responsibility of the user. We recommend that anyone intending to rely on any recommendation of materials or procedures for adhesive bonded structural repair mentioned in this publication should satisfy himself as to such suitability, and that he can meet all applicable safety and health standards. Methods and materials discussed may be potentially hazardous. **"Cautions," "warnings," and "special notes" printed in the text should be heeded.**

The book is for information only. It is strongly recommended that users seek and adhere to the manufacturer's or supplier's current instructions for handling each material they use, and obtain expert advice before implementation of any procedures.

CONTENTS

1

INTRODUCTION

The purpose of this handbook is (1) to provide a standard method for repairing of adhesive bonded and composite laminate structures, (2) to identify suitable materials and equipment for making satisfactory repairs, and (3) to discuss methods by which structures may be inspected, both before and after repair. The objective in selecting the methods and materials is to restore the structure to its original integrity, i.e., to make a permanent repair. The use of these methods is recommended to promote standardized repair procedures.

SCOPE AND LIMITATIONS

This handbook contains descriptions of methods of repair for adhesive bonded structures. The types of structures include metal and composite faced sandwich structures and laminated reinforced composite structures. The items covered include pre-repair damage evaluation, preparation of the part for repair, description of the materials of repair, repair procedures and post-repair inspection. While certain materials are recommended in this handbook, this does not signify that these are the only materials which are capable of being used to repair the structure. Other materials may have equal or superior qualities when tested to a specific requirement. The materials that have been included are for information purposes. These materials are considered satisfactory for the intended purposes.

USAGE OF THE HANDBOOK

The general procedures for the use of this handbook are as follows:

1. Determine the size or extent of the damage. Procedures to be used for damage evaluation are covered under Damage Assessment and are described in more detail under Nondestructive Inspection.
2. Determine the type of structure, i.e., sandwich or laminated structure.
3. Review the individual instruction manual for the structure in question for specific information and instructions for the structure being repaired. These typically include the following:
    - o Structural criticality of the damage area.
    - o Limitations, such as the maximum weight, that the repair may add to the structure, etc.
    - o Identification of the materials used in construction of the damaged part and the service temperature requirements.
    - o Limitations on the type of repair that is recommended depending on the location and type of repair. If no restriction are placed upon the repair of the structure the repairs in this handbook will generally be satisfactory.
4. Determine the appropriate repair method. Any limitations on the use of these methods will be the responsibility of the using agency. The methods of repair are given in the appropriate repair chapter.
5. Obtain the materials and equipment necessary to make the repair. These are described under Repair Materials.

6. Prepare the area for repair as described in the appropriate method. Pay particular attention to caution notes.
7. Observe the instructions for storage and handling of repair materials. Refinish and inspect the repair as required. Inspection techniques are presented under Nondestructive Inspection.
8. In various sections of this handbook reference is made to review by the cognizant engineering agency. Each user of this handbook should identify this agency, whether it be local engineering, the manufacturer or another organization.

PERSONNEL QUALIFICATION

Repair personnel should be fully qualified in the preparation of the repair materials and in the repair of structures. Inspectors must be qualified in the use of the inspection equipment required and should be skilled in the repair procedures. Qualification requirements for bonding technicians are covered in MIL-A-83377 (Ref. 1-1). Qualification of inspection personnel is covered in MIL-STD-410D (Ref 1 - 2).

WORKMANSHIP

The ability of the bonded repair to perform satisfactorily for the remaining life of the structure is very dependent on how well the repair is made. It is essential that the procedures outlined in this handbook are carefully followed and that all repairs are done in a good workmanship manner. Honeycomb structure can be extremely fragile. Special care is required to prevent damage to the surrounding structure during the repair operation. Protective coverings should be placed around the repair area where applicable.

REPAIR ATMOSPHERIC ENVIRONMENT

The requirements for the adhesive bonding environment are specified in Military Specification MIL-A-83377. The areas where the surface preparation of metal and composite parts are done should be isolated from operations that generate oil or water vapors or other contaminants. Composites should be machined in an environment which will remove the dust particles generated, such as a down-draft work station. All personnel handling cleaned parts must wear clean, white lint-free gloves.

Immediately after cleaning, parts should be moved into a controlled atmosphere area for bonding assembly. If the cleaning and controlled bonding areas are not in the same proximity, after cleaning, the parts should be sealed in non-contaminating wrapping for transfer to the controlled atmosphere layup area. The environment in the controlled area shall be as specified in MIL- A-83377. Operations that generate dust or other airborne contaminants, such as sanding and grinding must be prohibited in the controlled area. Similarly, smoking and eating in the controlled area should be prohibited.

Where practical, it is recommended that parts be removed from the main structure for repair in the shop. It is recognized, however, that some repairs will have to be made on the main structure. In these cases, special care must be taken to protect the surrounding surfaces from damage. Bonding of the part should occur as soon as possible after preparation of the repair area and the replacement part has been completed.

REPAIR MATERIALS

All materials used in bonded repairs shall conform to applicable specifications. Care must be taken to ensure that materials requiring refrigeration are maintained at the proper temperature during transit and placed in receiving storage.

During storage, proper temperature and humidity controls must be provided and maintained.

Upon receipt, it must be ensured that materials meet minimum specification requirements. This can be accomplished by performing in-house tests or by requiring the supplier to furnish certified qualifying test results. A guide to the type of tests that should be conducted in-house is discussed under Adhesive Receiving Acceptance Tests. All material should be clearly marked to indicate its storage and expiration dates. Outdated material must be requalified per the applicable specification or disposed of in an acceptable manner.

Instructions regarding material handling and storage must be followed. See Chapter 2 for detailed information on the storage and handling of adhesives and prepregs. The materials should be hermetically sealed before being returned to storage and the out-time recorded. Special precautions should be taken in the handling of flammable materials.

QUALITY ASSURANCE

An effective quality assurance program should be conducted concurrent with the repair procedures to assure satisfactory end-item strength and durability. Materials and material handling should meet requirements in the applicable specifications including proper storage and the adherence to shelf life stipulations. Processing steps should be carefully followed with emphasis on maintenance of properly prepared surfaces and good prefit of detailed parts. Processing procedures should be checked at critical steps to assure conformance. After completion, the quality of the repair should be evaluated by the responsible Quality Assurance personnel.

SAFETY PRECAUTIONS

The following safety precautions must be strictly observed while making repairs on a structure.

1. If the repair is to be made while the component is on an aircraft or similar structure, the structure and repair cart, if used, must be statically grounded. Only approved explosion proof electrical equipment should be used. Electrical equipment must be grounded while in operation.

2. When repairs are to be made over a fuel tank area, the fuel tank needs to be purged and checked continuously for an explosive mixture. The repair area should be kept well ventilated.

   Special precautions should be taken while working with flammable materials. Fire fighting equipment must be available during the repair operation.

3. Adequate ventilation is needed during the mixing and use of adhesives, sealants and solvents. Avoid breathing fumes from these materials.

4. When mixing acid solutions always add the acid to the water with agitation. Never add water to an acid.

5. Acid and alkaline solutions should not come in contact with the skin or clothing. In case of contact, they should be washed off immediately with generous amounts of cold water. Always wear eye protection and rubber gloves when using these solutions.

6. Wear heat insulating gloves when handling hot equipment and materials.

7. Respirators should be worn for any operations creating excessive dust such as sanding, grinding and the machining of metals and composites.

8. **Closely Observe All Applicable Local And Federal Safety Standards And Regulations.**

Special Note Concerning The Use Of Solvents

Solvents present potential health and fire hazards that must be given special consideration. Pertinent general information concerning the use of solvents is given in the subsequent paragraphs. Refer to other documents such as OSHA, EPA, and military regulations and standards.

Health Hazards:

1. If absorbed through the skin, solvents may cause dermatitis. They can dissolve natural skin oils and result in drying and cracking of the skin, rendering it susceptible to infection. Solvents may cause irritation and allergic reactions to sensitive individuals.
2. If vapors are inhaled, solvents can cause mild symptoms of headache, fatigue, nausea, or visual and mental disturbances under prolonged and repeated exposures to moderate concentrations. Severe exposure may result in unconsciousness and even death. Solvent vapors can also act as an anesthetic, or cause irritation of the eyes and the respiratory system. They can result in blood, liver and kidney damage.
3. Solvents are harmful if swallowed. Symptoms may be similar to those of vapor inhalation.

Minimize Personal Exposure:

Personal contact with the liquid or the inhalation of vapors should be minimized or eliminated by engineering techniques such as enclosing the process or equipment, isolating operations and using local exhaust ventilation and protective clothing and equipment.

Personnel should:

1. Avoid solvent contact with the skin. Wear rubber or neoprene gloves when handling liquid solvents. Other equipment, such as impervious aprons, sleeves, and coveralls may be necessary in certain operations.
2. Avoid eye exposure to liquid solvent, vapors or over-spray

by wearing chemical goggles or other approved eye protection.

3. Avoid using solvents as skin cleansing agents. If solvent contacts the skin, wash the affected area immediately with soap and water and apply a skin conditioning cream, lotion or ointment.
4. Avoid breathing solvent vapors. Use solvents only in a well ventilated area. Use respirators such as the chemical cartridge type gas masks or airline full-face respirators where there is a lack of engineering control and high vapor concentrations exist.
5. Avoid using solvents for unauthorized or unapproved purposes. Use only for purposes called out in the appropriate specifications or procedures.

Minimize Fire Hazards:

To eliminate or minimize the danger of fire and consequent destruction of life and property, flammable solvents should be used only in approved areas and with methods recommended by the local fire safety authority. These include:

1. All flames, smoking, sparks and other sources of ignition must be eliminated from areas using solvents.
2. Non-spark producing tools should be used.
3. Clothing or processes creating static electricity should be eliminated or properly grounded.
4. All electrical equipment (lights, motors, wiring, etc.) must meet the electrical and fire codes for such locations.
5. Flammable solvents should be in closed containers and only in quantities to satisfy immediate use.
6. Provide adequate ventilation to prevent buildup of vapors.

Special Notes Concerning Dust Generated By Machining Composites

The composite materials of concern are defined as fibrous material systems consisting of two or more constituents. Generally one constituent acts as a reinforcing agent (fibers of glass, Kevlar or graphite) and the other serves as the matrix or binder (epoxy, polyester, or other polymeric resin) These materials, when sanded, ground, drilled or machined can produce a dust which is potentially dangerous to health and, can also produce an explosive mixture.

Health Hazards:

1. If inhaled the dust produced by machining composite materials may cause permanent damage to the respiratory system.
2. Contact with the eyes may cause serious eye damage.
3. Contact with the skin may cause irritation which could lead to infections.

Minimize Personal Exposure:

1. Avoid breathing of dust. Work in a well ventilated (down-draft) work area. Use approved dust respirator.
2. Avoid eye exposure to the dust. Use approved eye protection.
3. Avoid contact between dust and skin. Use gloves, protective sleeves and other approved equipment.

Minimize Fire Hazards:

To eliminate or minimize the danger of fire and explosions from dust, the machining of composite materials should be done in areas approved and with methods recommended by the local fire safety authority. These include:

1. All flames, smoking, sparks and other sources of ignition must be eliminated from the work area.

2. Dust particles should be removed from the work area by a properly designed down-draft work station.
3. Clothing and/or processes creating static electricity should be eliminated or properly grounded.
4. All electrical equipment (lights, motors, wiring, etc.) must meet the electric and fire codes for such locations.

Special Note Concerning The Use of Chemicals

The use of chemical materials requires special handling and record keeping. All incoming materials require a Material Safety Data Sheet (MSDS). This MSDS supplies information about the make-up of the material, identification of hazardous ingredients, fire and explosion data, health hazard information, reactivity data, disposal, spill and leak procedures, and special protection information. All-on-site chemical baths etc. must be properly controlled and monitored. Disposal of hazardous materials must be controlled, monitored and documented. See local, state and federal regulations pertaining to the control and disposal of chemical materials.

REFERENCES

1 - 1 Military Specification MIL-A-83377: "Adhesive Bonding (Structural) for Aerospace and Other Systems, Requirements for"

1 - 2 Military Standard MIL-STD-410D, "Qualification of Inspection Personnel"

2

MATERIALS AND PROCESSING

MATERIAL SELECTION

This chapter contains information pertinent to the selection of materials used for repairing adhesive bonded structures. The materials, classified by general usage, are as follows:

- Adhesive, Repair, Paste
- Adhesive, Laminating
- Adhesive, Structural Film
- Adhesive, Core Splice
- Adhesive, Injection
- Adhesive, Aerodynamic Smoother
- Sealant/Aerodynamic Smoother
- Prepreg
- Laminates
- Core Materials

The materials listed in the tables in this chapter are intended as guides in making repairs. No restrictions are placed on the use of equally acceptable materials except that their use must be approved by the applicable authority.

DESCRIPTION OF MATERIALS

Adhesive, Repair, Paste

The paste type repair adhesives are generally two part thixotropic pastes having virtually no sag. These adhesives are capable of curing at room temperature, however, the preferred cure may be at an elevated temperature. These adhesives have good strengths and they develop durable bonds.

Adhesive, Laminating

The laminating adhesives are two part, low viscosity liquid systems designed to impregnate and wet-out the fiber material used to make the composite. These adhesives are capable of curing at room temperature or at a moderately elevated temperature.

Adhesive, Structural Film

The structural film adhesives are adhesives that have been fabricated in film form and generally are supported by a woven or mat carrier often called a scrim. They are one part systems which have the curing system already incorporated into the adhesive. They require heat and pressure for a full and useful cure. These adhesives require refrigerated storage until they are to be used (see handling Storage and Handling of Adhesives).

Adhesive, Core Splice

The core splice adhesives are adhesives that have been specially designed to bond the edges of honeycomb core together or to an edge member. They accomplish this by expanding at least 200% during cure, thus filling in the open cell areas at the edge of the core (Figure 2-1). These systems require elevated temperature cures and are often cured at the same time as the structural film adhesive, provided that the cure cycles are compatible. When a core splice adhesive is not available, multi-layers of film adhesive may be used as a substitute, however this will add extra weight. Core splice pastes are also available.

Figure 2-1. Expanded core splice adhesive in use to bond core. (Courtesy of McCann Manufacturing Inc.)

Adhesive, Injection

The injection type adhesives are low viscosity, two-part, room temperature curing systems that can be injected through a needle into a void. These systems can also be cured at elevated temperatures.

Adhesive, Aerodynamic Smoother

The aerodynamic smoother adhesives are two part systems that are thixotropic. Their purpose is to fill nicks, dents and crevasses in a surface. They are machinable so as to produce an aerodynamic smooth surface. They are capable of staying in place during service without an external cover other than paint.

Sealant/Aerodynamic Smoother

A sealant is a polymeric material that is used primarily to provide isolation from an environment. The sealant may be used as a moisture barrier, a fuel/solvent barrier, a differential pressure barrier, an insulator or for aerodynamic smoothing. The sealants used for repair are dependent upon the service temperature, fuel/fluid resistance criteria, and in some cases an aerodynamic smoother requirement. Corrosion-inhibiting and non-corrosion resistant versions are available for service temperatures up to 350 degrees F. These systems are generally two-part systems and are available in variable working times.

Prepregs

A prepreg material is a resin preimpregnated fiber carrier which when cured, either alone or in layers, will form a product called a laminate. The resin used in fabricating the prepreg is in a B-staged state and, therefore, must be stored under refrigeration (see Storage and Handling of Adhesives & Prepregs). Prepreg materials are cured under heat and pressure.

Laminates

A laminate is a cured resin impregnated fiber material. The number of layers or plies may vary from two to as many as the design requires.

Core Materials

Core materials vary depending upon the structure. The

more frequently used core materials include honeycomb core, balsa wood, foams, plywood and some of the newer structural plastics as well as some special designed materials. The core material is used in structures to stiffen thin skins and to allow the skins to carry higher compressive loads. The core material used in a repair should be the same as was used in the original construction. If honeycomb core is used, the density cell size and ribbon direction of the original core must be maintained. These materials will be discussed in detail in subsequent paragraphs.

REPAIR MATERIALS

This section covers the materials of repair, the suggested cures for repair adhesives and sealants, suggested operating conditions for machining composites containing Kevlar, and a list of material and equipment suppliers.

The materials of repair are given in Table 2-1. The table lists the materials of repair, the supplier of the material, description or manufacturer's identification number, federal stock number or specification number if available, and general remarks.

Table 2-2 contains the recommended cure cycles for the repair adhesives and sealants. Kevlar containing composites present special problems when they are to be machined, cut, drilled, etc. The suggested operating conditions for machining composites containing Kevlar are listed in Table 2-3. This information was obtained from "Cutting and Machining Kevlar" by DuPont.

Table 2-4 lists the materials and equipment supplier for items listed in Table 2-1.

Table 2.1 Materials of Repair

Material	Supplier	Description (Mfgr's No.)	Remarks
ADHESIVES Repair	Dexter/Hysol Aerospace Prod. ATACS Prod,Inc.	EA 9394 5146	2 part paste 2 part paste
Laminating	Dexter/Hysol ATACS	EA 9396 5138	2 part paste 2 part paste
Aerodynamic smoother	ATACS Dexter/Hysol 3M	5146 or 5103 EA 9309.2 EC 2216	2 part paste 2 part paste 2 part paste
Structural	Dexter/Hysol BASF Narmco 3M McCann Mfg. ATACS	EA 9628 M 1113 AF 163 MA 429 0617	supported film supported film supported film supported film by linear yard
Core Splice	McCann Mfg. Cyanamid ATACS	MA 562 FSN8040-00-185-5631 FM 410 series 0605	expanding film expanding film expanding film
Injection	Dexter/Hysol ATACS	EA 9309.2 5138	2 part paste 2 part paste
MATERIALS Abrasive disc	any source	100, 400 grit	
Abrasive disc holder	Standard Abrasive Inc.	SM 3-T-4A pad/ stem	
Abrasive paper	any source	100, 400 grit	
Drill, cobalt	any source	135 degree point angle	for drilling titanium
Drill, hi-speed	any source	90/108 degree angle	for drilling aluminum
Drill stop spring loaded	Monogram Aerospace Fasteners	Wedgelock DS 10, 20, 30 & 40	for setting drill depth

File, hand rotary	any source	1/2, 3/4, or 1 inch diameter	for removal of metal and honeycomb core
Gloves, cotton	Cardinal Glove Company		FSN 8415-00-268-8353
Heater blanket	ATACS Products Briscoe Mfg. Co.	7000 series	Custom sizes or shapes 115 V, 60 cycle, 5-10 W/sq. inch not to exceed 30 amps/zone
Knife, putty		1 1/4 inches wide	FSN 5120-294-4605
Knife, retractable blade			FSM knife 5110-892-5071 FSM blade 5110-293-2865, Type 1
Methyl-ethyl-ketone	any source		FSM 9G-6810-281-1785 ASTM D740
Needle, injection	Semco-Bancroft Philip Fishman	FN-13G	Stainless steel
Needle, adapter	Semco-Bancroft Philip Fishman	F-2 F-3	1/4 pipe thread for Semco cartridges for syringes
Pliers, duckbill 6 inch			FSN 5120-256-2150 Fed GGG-P-471
Pliers, needle nose, straight			FSN 5120-256-2150 Fed GGG-P-471
Router, attachments	Dresser Inds Tool Group	QRC-3C or 3D Collet-9003 Nose-9013, 9022 Bearing-9015 Ret. ring-9016 #881390 Complete	use with 3C or 3D router motors use with 11GLF 230/250 router motor
Router bits		1/4 inch diam, 2 flute LH spiral, hi-speed 1/4 inch diam, 3 flute LH spiral, hi-speed	for aluminum for titanium

Router bushing	Sheridan Products Inc.	SP-310-31J	use with QRC 3C and 3D router motors
Router motor	Dresser Inds.	QRC-3C or 3D 20,000 rpm 11GLF-230/250 25,000 rpm	use with QRC attachments use with #881390 attachment
Release agent	Ram Chem Inc. Dexter/Frekote 33 Airtech Intl. ATACS products	Garan 225 Frekote 33 Release All-1000 7205	
Release films	Airtech Int. ATACS Products Dupont	E 3630 TFE 0502-3 TFE 0501-3 TFE FEP or TFE	0.0015 or 0.002 inch thick non-porous porous 0.0015 or 0.002 inch thick
Saw, circular with adapter and blade	Chicago Pneumatic Tool Co.	CP-3017-OKS-1000-3	for cutting panel skins
Saw, hole	any source	1/2, 1, & 2 inch diameter	for use in damage removal
Scribe	any source	reversible point	for detail layout
Sealants	PRC PRC	Proseal 890C PR 1432G PR 1422G PR 1436G	FSN 8080-723-2746 2 part general purpose 2 part Corrosion inhib. chromate type
Spatula	any source	wood, plastic or metal	for mixing and spreading of adhesives
Syringe	Semco-Bancroft	E-604, 35cc	use with F3 adapter
Tape, aluminum faced	Permacel 3MM	P-12 Y-9301	
Tape, masking (high temp)	Borden Chemical	Mystic 6325	AA 883

Temperature controller, console portable self-contained	ATACS Products	8010 single zone 8020 two zone	FSN
Trichloro-methane	any source		O-T-620
Vacuum bagging film	Airtech Intl. ATACS Products DuPont DuPont DuPont	8171 0551 Tedlar 200	Nylon Nylon Nylon Mylar Polyvinyl fluoride
Vacuum pump	ATACS Products Airvac Engrg.	4903 TD 206	air aspirator air aspirator
Vacuum bag quick disconnect	Airtech Intl. ATACS Products	AQD 500BF 4903	
Vacuum bag valve	Airtech Intl. ATACS Products	401 4909	
Vacuum bag valve base plate	Airtech Intl. ATACS Products	401 4909	
Vacuum bleeder	Airtech Intl. ATACS Products	Airweave SS 0507	
Vacuum sealing	Airtech Intl. ATACS Products General Sealants	GS 213 2403 GS 213 GS43	Putty, non-curing, zinc chromate
Abrasive discs & finishing discs	3M	Scotch Brite Silicone Carbide paper	60-240 grit 80-100 grit, use wet
Counter sinks	Airtech Intl. ATI Inds. Guehring Intl. Carbide Weldon tools		

Drill bit, brad point	Airtech Intl. Pen Assoc. Starlite Inds. VNE Corp. A&W Tool & Machine Ltd.		
Drill bit, self-centering	Penn Assoc.		
Drill bit, single fluted	Starlite Inds.		
Drill bit, spade	Penn Assoc. Technology Assoc.		
Edge sander, power	Dotco Dynabrade Nitto Kohki USA Penn Assoc.		
Grinders	any source	Alumina	
Knibbler	Atlas Copco ATI Inds.		for cutting thin laminates
Mill cutters	Penn Assoc.		
Power shears	Airtech Intl.		for cutting cloth and prepregs
Router bits, conventional	Starlite Inds. Penn Assoc. Spacematic cutting tools Onsrud Cutter Ultra Tool Intl. Randell Midwest		require secondary edge finishing
Router bits, split helix	Guehring Airtech Intl. Technology Assoc.		
Saw, band carbide grit	Remington Abrasive Pdts.		
Saw, band conventional	Pen Assoc.		

Saw, circular	Pen Assoc. Remington Abrasive Pdts. Technology Assoc.		
Saw, saber	Airtech Int. Pen Assoc. Technology Assoc.		
Tungsten carbide razor blades	Pen Assoc.		
Ultrasonic knife	TMI Ultrasonic Tool Inc. Branso Sonic Power		for cutting kevlar cloth and prepregs

Table 2.2 Recommended Cure Cycles for Repair Adhesives and Sealants

Material	Description	Primary Cure	Alternate Cure
Adhesive	EA 9394 ATACS 5146 EA 9396 ATACS 5138 ATACS 5103 ATACS 5146 EA 9309.2 EC2216 EA 9628 M1113 AF 163 MA 429 ATACS 0617 MA 562 FM 410 ATACS 0605	1 hr @ 150F 3 hr @ 150F 1 hr @ 180F 3 hr @ 150F RT 15 min 1 hr @ 150F 1 hr @ 150F 2 hr @ 150F 1 hr @ 250F 1 hr @ 260F 1 hr @ 250F 1 hr @ 250F 1 hr @ 250F 1 hr @ 250F 1 hr @ 250F 1 hr @ 250F	RT 7 days RT 7 days RT 7 days RT 7 days --------- RT 7 days RT 3 days 90 min @ 225F 120 min @ 200F 90 min @ 225F 90 min @ 225F 90 min @ 225F 90 min @ 225F 90 min @ 225F 90 min @ 225F
Sealants	Proseal 890 PR 1422G B-2 PR 1432GP PR 1436G	RT 14 days RT 3 days RT 3 days RT 3 days	

Table 2-3 Suggested Operating Conditions For Matching Composites Containing Kevlar[1]

Tool Description	Suggested Speed	Feed Rate	Remarks
Single fluted drill	200-600 SFPM	0.002-0.005"/rev	Drill locator bushing required Backup support required
Brad point drill	6000-25000 RPM	0.002-0.005"/rev	Controlled feed rate required Cannot be used to drill metals
Spade drill	250-450 SFPM	0.005-0.010"/rev	Special operational techniques required
Hole saw	150-300 SFPM	0.005-0.010"/rev	Backup required
Counter sinks	50-250 RPM		Micro stop tooling required
Band saw conventional	3000-6000 SFPM	up to 36"/Min*	Raker or straight set
Band saw carbide grit	3000-6000 SFPM	up to 36"/min*	Use with Hybrid laminates
Circular saw	3000-6000 SFPM	up to 36"/Min*	Coolant** may be required for heavy cuts
Saber saw	2500 strokes/ Min	up to 36"/Min*	
Router bits	20,000-27,000 RPM	up to 60"/Min*	Some secondary edge finishing may be required
Edge sanding	4000-4000 SFPM		80-180 grit aluminum oxide or silicon carbide. Belt sanding preferred over disk sanding.
Power shears		30-40"/Min*	0.1" thick max material

Notes on machining Kevlar:

1. Data obtained from "Cutting and Machining Kevlar" U.S. Source List. Manufacturers/Vendors of Tools for Cutting & Machining Composites of Kevlar
* Depends on material thickness.
** Suggest water soluble coolant: 150-200 parts water/1 part coolant concentrate.

Table 2-4 Material and Equipment Suppliers

Supplier	Address
Airtech International	2542 East Del Amo Blvd. P. O. Box 6207 Carson, CA 90749 213-603-9683
Air Vac Engineering Co.	Milford, CT
ATACS Products Inc.	1120 SW 16th St. Renton, WA 98055-2909 206-226-8340
ATI Industries	2425 West Vineyard Ave. Escondido, CA 92025-2591 619-746-8301
Atlas Copco	24404 Indo Plex Circle Farmington Hills, MI 48019 619-521-8301

Supplier	Address
A & W Tool & Machine Ltd	P. O. Box 395 2326 York Rd. Timonium, MD 21093 301-252-2200
BASF/Narmco Products	1440 N. Kramer Blvd. Anaheim, CA 92806 714-630-9400
Borden Chemical Co.	Mystic Tape Div.
Briscoe Mfg. Co.	1055 Gibbard Ave. P.O. Box 628R Columbus, OH 43216
Cardinal Gloves Co.	1 Lisbon St. Clifton, NJ 07013 201-472-8787
Chicago Pneumatic Tool Co.	6 East 44th St. New York, NY 10017
Cyanamid	American Cyanamid Co. Old Post Rd. Havre deGrace, MD 21078 301-939-1910
Dexter Adhesives & Structural Matls Div. (Frekote Products)	164 Folly Mill Rd. Seabrook, NH 03874 603-474-5541
Dexter Adhesives & Structural Matls Div. Hysol Aerospace Products	2850 Willow Pass Rd. P. O. Box 312 Pittsburgh, CA 94565-3299 603-474-5541

Dotco	Ohio Route 18E Hicksville, OH 43526 419-542-7711
Dresser Industries	Tool Group 6114 6th Avenue South Seattle, WA 98108
Dynabrade Inc.	72 E. Niagara St. Tonawanda, NY 14150 715-694-4600
E.I. DuPont DeNemours & Co. Inc.	Wilmington, DE 19898
General Sealants	15248 Proctor City of Industry, CA
Guehring	1455 Commerce Avenue Brookfield, WI 53005 414-784-6730
International Carbide	1111 N. Main St. Wauconda, IL 60084 800-323-7440
McCann Manufacturing Co.	P. O. Box 429 Route 14A Oneco, CT 06373 800-331-1353 203-564-4046
Monogram Aerospace Fasteners	2343 So. Garfield Ave. Los Angeles, CA 90040

Morris Tool Ltd.	G-4006 Corunna Rd. Flint, MI 48504-5891 313-732-3550
3M ACS Division	223-1N-07 3M Center St. Paul, MN 55144-1000 612-733-1441
Nitto Kohki USA	111 Charlotte Place Englewood Cliffs, NJ 07632 201-568-7980
Onsrud Cutter	800 Liberty Dr. Libertyville, IL 60048 312-362-1560
Permacel	P. O. Box 671 New Brunswick, NJ 08903 201-418-2400
Pen Associates	2639 W. Robino Dr. Wilmington, DE 19808 302-995-6868
Philip Fishman Co. Inc.	7 Cameron St. Wellesley, MA
Products Research Corp.	5454 San Fernando Rd. Glendale, CA 91203 818-240-2060
Ram Chemical Inc.	Gardena, CA
Randell Midwest Co.	9741 Barnum Ave. Bloomington, MN 55431 612-881-7997

Remington Abrasive Pdts.	939 Barnum Ave. Bridgeport, CT 06601 203-333-1112
Semco-Bancroft	Rock Hill Rd. Cherry Hill, NJ 08003 609-854-8000
Sheridan Products Inc.	1054 E. Hyde Park Blvd. Inglewood, CA 90302
Spacematic Cutting Tools	138 Hulls Farm Rd. Southport, CT 06490 203-254-1863
Standard Abrasive Inc.	19015 Parthenic St. North Ridge, CA 91324
Starlite Industries	1111 Lancaster Ave. Rosemont, PA 19010 215-527-1300
Technology Associates	P. O. Box 7163 Wilmington, DE 19803 302-475-6219
Ultrasonic Tool Inc.	P. O. Box 389 964 E. Swedsford Rd. Exton, PA 19341 215-644-2620
Ultrasonic Tool International	5451 McFadden Ave. Huntington Beach, CA 92649 800-854-2431

VNE Corp.	P. O. Box 187 1415 Johnson Street Janesville, WI 53547 800-356-1111
Weldon Tool Co.	3002 Woodhill Rd. Cleveland, OH 44104 216-721-5454

STORAGE AND HANDLING OF ADHESIVES AND PREPREGS

Storage of Adhesives

The room temperature curing paste type adhesives can generally be stored at ambient temperature, provided that ambient is not much over 77 degrees F (25 degrees C). The shelf or storage life of the adhesives can be extended by cold storage. Storage at 40 degrees F will often double the storage life, and storage at 0 degrees F could double it again. Most film or elevated temperature curing adhesives require storage at temperatures below zero degrees F. These systems will generally have an out time (time it can be left at room temperature) that is measured in days.

Storage of Prepregs

Most prepregs can be described as a combination of a film adhesive and one or more layers of a composite material which have been layered by some propriety process. Therefore, the storage life is very similar to a film adhesive, and cold storage is required.

Handling of Adhesives

Paste adhesives that are stored under refrigerated conditions should be allowed to equilibrate to room temperature before the container is opened to prevent condensation of moisture on the adhesive. Film adhesives should be removed from cold storage at least 12 hours prior to use to allow stabilization at room temperature. Keep the roll wrapped in its plastic storage bag to prevent moisture from condensing on the adhesive. All adhesives should be handled with care. Avoid skin contact as some ingredients can cause skin irritation. The use of clean lint free gloves is recommended both from the hygienic standpoint and for the reliability of the repaired structure.

Handling of Prepregs

The requirements for handling prepreg materials are the same as for handling film adhesives.

MIXING AND CURING REQUIREMENTS FOR ADHESIVES AND SEALANTS

Check the manufacturer's data sheet for the work life of the mixed material (this may be given as pot life). Note the size of the batch listed and mix only the amount of material that can be used within the specified time. Cure cycles are given in Table 2-2. Specific applications may require deviations. In those cases, the manufacturer's recommendations for mixing, heat-up rates, cure time and pressure requirements should be followed.

ADHESIVE RECEIVING ACCEPTANCE TESTS

It is recommended that adhesives used for the repair of adhesive bonded structures should be tested by the user to ensure compliance with the specifications. If this is not practical a certificate of compliance should be obtained from the supplier. The acceptance tests should include the following:

a. Lap shear strength per ASTM D1002 or ASTM D3165. Lap shear tests should be performed at room temperature and at the highest use temperature, e.g., 180 degrees F for a 250 degree F curing adhesive.

b. Peel strength per ASTM D3167. Peel tests should be performed at room temperature and at -65 degrees F.

c. Sandwich peel strength per ASTM D1781.

d. Film weight (for structural film adhesives). To determine the film weight, cut 2 specimens of material across the width of the film, approximately 6 x 6 inches. Determine the area to the nearest 0.1 square inch. Determine the weight of the film minus the separator sheets to the nearest 0.1 grams. Calculate and report the average weights to the nearest 0.001 lbs/sq. ft.

In addition to the tests specified above any other test described applicable specification may be used to ensure that the adhesive conforms to specification and are comparable to material previously qualified.

MATERIAL PROCESSING

Application of Paste Adhesives

1. Allow any adhesive removed from refrigerated storage to reach room temperature prior to opening the container to prevent moisture condensation.

2. Weigh the amount of material required for the application. NOTE: Mix only the amount of material that can be used within the specified working life. A large quantity of mixed material will shorten the working life (exothermic reaction) and hasten gel time.

3. With preweighed or premixed component materials, follow the manufacturer's mixing and application instructions.
4. Mix to a uniform color. Mix slowly to minimize air entrapment.
5. Apply a uniform coating to each fading surface by trowel, brush, roller, etc.
6. Assemble the details.
7. Secure in place.
8. Bag and apply pressure as described in Vacuum Bagging of Structure Under Repair.
9. Cure as specified in Table 2-2 or as specified by the manufacturer.

Application of Film Adhesives

1. Remove from refrigerated storage and allow to equilibrate at room temperature before opening the wrapper. Record the adhesive "out time" and keep a cumulative record. Requalify or discard when the allowable "out time" has been reached.
2. Solvent clean cutting tools, templates, and other equipment used for laying out the adhesive.

   CAUTION: Do not touch the adhesive with bare hands or other parts of the body. Clean, white, lint free gloves must be worn when handling prepared details, adhesives or other parts during the lay-up.
3. Apply a layer of adhesive film to one of the faying surfaces. Do not fold, stretch, or otherwise thin the adhesive while handling.
4. Leave the separator sheet on the side of the adhesive exposed to the atmosphere as a temporary protective cover.
5. Press the adhesive film smoothly in place. Avoid air entrapment.
6. Trim the adhesive film to leave approximately 1/8 inch

excess beyond the joint perimeter. The adhesive film must not be less than net, i.e., not less than the size of the part. The details may be used as templates for cutting the adhesive.

7. Remove the remaining separator sheet just prior to assembly of the details.
8. Prepare for cure as specified in Vacuum Bagging of Structures.

Under Repair Application of a Laminating Adhesive

1. Follow the mixing instructions given under Application of Paste Adhesive.
2. Apply a small amount of adhesive to the fabric and work it into the fabric with a squeegee, until the fabric is saturated, working out all entrapped air. Place the fabric in the required position. Repeat until the required number of plies have been positioned.
3. Apply a sheet of perforated release film over the part. Cover the release film with a bleeder cloth and a caul plate. Bag as described in Vacuum Bagging of Structures Under Repair.
4. Cure per Table 2-2 or in accordance with the manufacturer's instructions.

Application of Core Splice Adhesive

1. Core splice adhesives are films similar to the structural film adhesives, therefore, they are be handled like film adhesives.
2. Cut the adhesive in strips. The width of the strips should equal the thickness of the core section to be bonded.
3. Remove the separator sheet from one side and press the adhesive firmly to the edge of the core. Remove the other separator sheet and press the next section of core to

the adhesive. Continue until all sections of core have been joined.

4. Continue with the operation of assembling the remainder of the parts, film adhesive, and skin.
5. Bag as per Vacuum Bagging of Structures Under Repair, and cure as per Table 2-2, or the film adhesive manufacturer's instructions.

Application of Sealants and Aerodynamic Smoothers

1. Allow refrigerated materials to reach room temperature before opening to prevent moisture condensation.
2. Prepare the applicable environmental sealant or aerodynamic smoother compound according to the manufacturer's instructions.
3. Observe safety precautions.
4. Remove adhesive flash at the edge of the bonded joint to be sealed as required. Leave approximately a 45 degree fillet for sealing the edges.
5. Wipe clean with solvent and clean cloth until clean cloth shows no sign of soil. Wipe dry with a clean towel or cloth.
6. Apply the sealant to all repair adhesive bond lines.
7. Cure the sealant or smoother per applicable instructions or specification.

It is important to note that sealant becomes more reactive at higher temperatures and higher humidities and less reactive at low temperatures and humidities. For example a material that has a work life of 4 hours at 60 degrees F and 35% RH may have a work life of 2 hours at 77 degrees F and 50% RH, and the same material may only have a work life of 1/2 hour at 90 degrees F and 65% RH.

SANDWICH CONSTRUCTION

Sandwich construction consists of generally three types of materials bonded together (Figure 2-2). The outermost laminations are called skins or facings and the inner lamination is called the core. The skins or facings of a panel act very much like the flanges of an I-beam by taking the bending loads. One skin is in tension while the other skin is in compression. The core material responds like the web of the I-beam. The core material resists the shear loads and increases the stiffness of the structure by spreading the skins apart. Furthermore, the core gives continuous support to the skins because the skin-to-core adhesive rigidly joins the component parts together into a single structure with high torsional rigidity. The skin loading is either shear, tension, compression, or a combination of tension or compression with shear (see Figure 2-3). The core loads are always shear, sometimes with compression or tension loads superimposed. Sandwich structure is strong, stiff, generally smooth, and lightweight. It may be flat, simple curved or compound curved. Typical uses of honeycomb structures are walls or bulkheads, floors, work decks or platforms, ramps, transverse and torsional shear webs, and many aircraft components. In the construction industry, doors and prefab wall panels are often fabricated using honeycomb construction. There are many uses for sandwich construction in the marine and transportation industries.

The core materials most frequently used in sandwich construction are honeycomb cores, foam cores and wood cores such as balsa wood. Some other types of wood are also used in some sandwich structures.

CORE MATERIALS

Honeycomb Core Materials

The most common or most widely used honeycomb core material is the aluminum honeycomb. However, other types of honeycomb core are available, including other metals such as

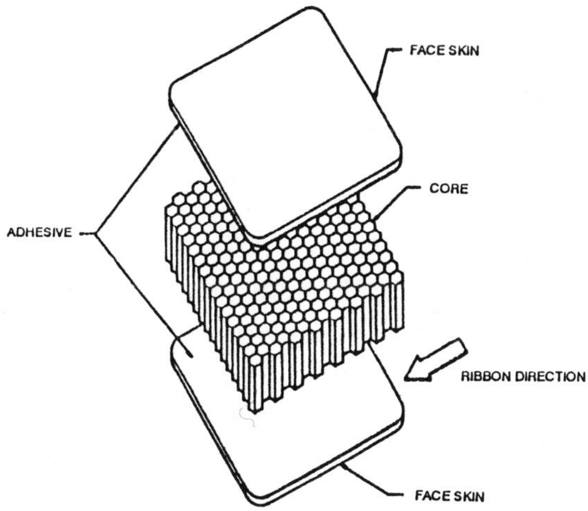

Figure 2-2. Typical sandwich construction.

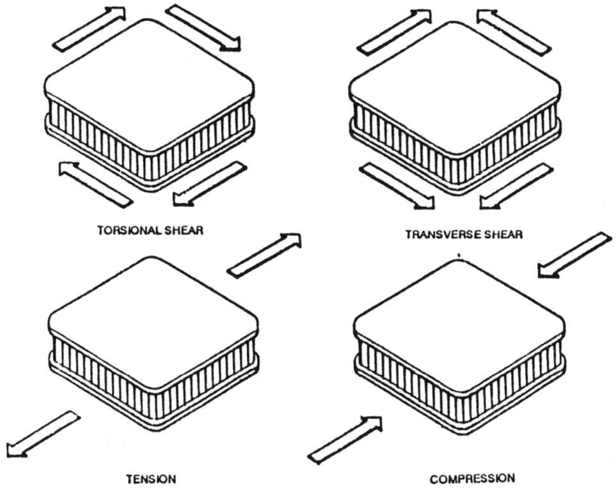

Figure 2-3. Sandwich construction skin loads

stainless steel, glass reinforced plastics, resin coated Aramid--fiber paper and the resin coated kraft paper.

The most common form of honeycomb is made by the "expansion" method. This method begins with the stacking of sheets of the web material on which adhesive node lines have been printed. The adhesive lines are then cured to form a HOBE (Honeycomb Before Expansion) block as shown in Figure 2-4. (Ref. 2-1)

The HOBE is cut into HOBE slices to the appropriate T dimension. Each HOBE slice is then expanded to the desired cell shape yielding an expanded panel. The expanded panels are trimmed to the desired L dimension (ribbon direction) and W dimension (transverse to the ribbon). An alternate method is to expand the cured HOBE to the desired cell shape resulting in an expanded block. Slices of the expanded block may then be cut to the desired T dimensions. High density honeycomb materials are generally made by the corrugated process as illustrated in Figure 2-5 (Ref. 2-1). In this process the web material is run through a set of corrugating rolls. Adhesive is applied to the corrugated nodes, and the sheets are

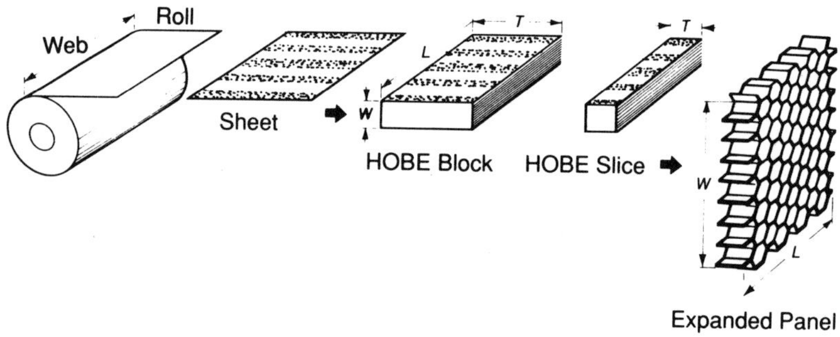

Figure 2-4. Expansion process of honeycomb manufacture. (Ref. 2-1)

then stacked into blocks and cured. Corrugated panels are then cut from these blocks to the required core thickness (T).

Metallic Honeycomb

The most common metallic honeycombs are made from 5052, 5056, or 2024 aluminum alloys to which a corrosion resistant coating has been applied before the node adhesive has

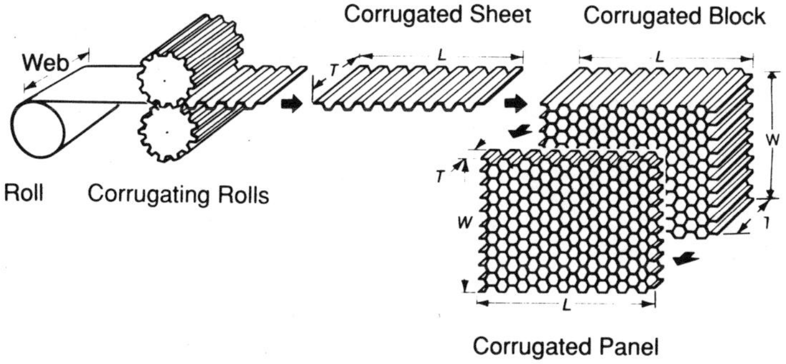

Figure 2-5. Corrugated process of honeycomb manufacture. (Ref. 2-1)

been applied, providing corrosion protection over the full foil surface. Hexcel also produces an ACG (Aluminum Commercial Grade) honeycomb which is a low cost aluminum honeycomb product for non MIL specification requirements. Special honeycombs such as stainless steel, titanium, or copper may be obtained in either adhesive bonded, welded, or brazed forms.

The properties of some metallic honeycomb core materials are given in Table 2-5 through Table 2-9 (Ref. 2-1 and 2-2)

Glass Reinforced Honeycomb

Glass reinforced honeycombs are made by first impregnating the glass web material with a resin material such as phenolic resin, a nylon-modified phenol, or a polyamide resin. The respective expanded panels are then dipped in a phenolic, polyester or polyamide resin. The properties of some glass reinforced honeycombs are given in Tables 2-10 through 2-13.

Aramid-Fiber Reinforced Honeycomb

Aramid-fiber reinforced honeycombs consist of DuPont's

NOMEX aramid-fiber paper treated with a heat resistant phenolic resin or polyamide resin. Properties of the aramide--reinforced honeycomb are given in Table 2-14. Honeycomb materials can also be made using DuPont's KEVLAR aramide--fiber paper.

Paper Honeycomb

Paper honeycomb materials are made of kraft paper and are impregnated and coated with a phenolic resin. This type of honeycomb core meets the requirements of MIL-H 21040

Table 2-5. Properties of Specification Grade 5052 Alloy Hexagonal Aluminum Honeycomb (Ref. 2-2)

Hexcel Honeycomb Designation / Cell — Material — Gage	Nominal Density pcf	Compressive					Crush Strength psi	Plate Shear					
		Bare		Stabilized				"L" Direction			"W" Direction		
		Strength psi		Strength psi		Mod ksi		Strength psi		Mod ksi	Strength psi		Mod ksi
		typ	min	typ	min	typ	typ	typ	min	typ	typ	min	typ
1/16—5052—.0007	6.5	870p	—	910p	—	275p	—	510p	480p	90p	320p	300p	40p
1/16—5052—.001	9.5	1500p	—	1550p	—	420x	—	850	835p	105x	540	520p	53
1/16—5052—.0015	12.0	2100p	—	2300p	—	650p	—	1150x	—	120x	670x	—	65x
1/8—5052—.0007	3.1	285	200	300	215	75	130	210	155	45.0	130	90	22.0
1/8—5052—.001	4.5	550	375	570	405	150	260	340	285	70.0	220	168	31.0
1/8—5052—.0015	6.1	1000	650	1040	680	240	450	550	455	98.0	320	272	41.0
1/8—5052—.002	8.1	1485	1000	1515	1100	350	750	725	670	135	455	400	54.0
1/8—5052—.003	12.0	2800	2100	3050	2200	900p	—	1940p	1250p	—	1430p	1000p	—
5/32—5052—.0007	2.6	220	150	260	160	55	90	165	120	37.0	100	70	19.0
5/32—5052—.001	3.8	395	285	410	300	110	185	270	215	56.0	175	125	26.4
5/32—5052—.0015	5.3	690	490	720	535	195	340	420	370	84.0	270	215	36.0
5/32—5052—.002	6.9	1080	770	1130	800	285	575	590	540	114	375	328	46.4
5/32—5052—.0025	8.4	1530	1070	1600	1180	370	800	760	690	140	475	420	56.0
3/16—5052—.0007	2.0	160	90	175	100	34	60	120	80	27.0	70	46	14.3
3/16—5052—.001	3.1	315	200	335	215	75	130	210	155	45.0	130	90	22.0
3/16—5052—.0015	4.4	560	360	590	385	145	250	330	280	68.0	215	160	30.0
3/16—5052—.002	5.7	820	560	860	600	220	390	460	410	90.0	300	244	38.5
3/16—5052—.0025	6.9	1120	770	1175	800	285	575	590	540	114	375	328	46.4
3/16—5052—.003	8.1	1660	1000	1720	1100	350	750	725	670	135	455	400	54.0
1/4—5052—.0007	1.6	90	60	100	70	20	40	85	60	21.0	50	32	11.0
1/4—5052—.001	2.3	210	120	230	130	45	75	140	100	32.0	85	57	16.2
1/4—5052—.0015	3.4	375	240	385	250	90	150	235	180	50.0	150	105	24.0
1/4—5052—.002	4.3	500	350	560	370	140	230	320	265	66.0	210	155	29.8
1/4—5052—.0025	5.2	720	500	760	510	190	335	410	360	82.0	265	200	35.4
1/4—5052—.003	6.0	1050	630	1100	660	235	430	495	445	96.0	315	365	40.5
1/4—5052—.004	7.9	1420	970	1490	1050	340	725	700	650	130	440	390	52.8
3/8—5052—.0007	1.0	45	20	50	20	10	25	45	32	12.0	30	20	7.0
3/8—5052—.001	1.6	90	60	95	70	20	40	85	60	21.0	50	32	11.0
3/8—5052—.0015	2.3	190	120	200	130	45	75	140	100	32.0	85	57	16.2
3/8—5052—.002	3.0	300	190	315	200	70	120	200	145	43.0	125	85	21.2
3/8—5052—.0025	3.7	405	270	420	285	105	180	260	200	55.0	170	115	26.0
3/8—5052—.003	4.2	620	335	640	355	135	220	310	255	65.0	200	150	29.0
3/8—5052—.004	5.4	810	500	850	535	200	360	430	380	86.0	280	228	36.8
3/8—5052—.005	6.5	1000	700	1100	750	265	505	545	500	105	350	300	43.5

Notes:
Test data obtained at 0.625 inch thickness.
p = Preliminaries
x = Predicted values.
I-Beam shear for 12.0 pcf product.

Table 2-6. Properties of Specification Grade 5056 Alloy Hexagonal Aluminum Honeycomb (Ref. 2-2)

Hexcel Honeycomb Designation Cell — Material — Gage	Nominal Density pcf	Compressive					Crush Strength psi	Plate Shear					
		Bare		Stabilized				"L" Direction			"W" Direction		
		Strength psi		Strength psi		Mod ksi		Strength psi		Mod ksi	Strength psi		Mod ksi
		typ	min	typ	min	typ	typ	typ	min	typ	typ	min	typ
1/16—5056—.0007	6.5	1000x	—	1100x	—	330x	—	645x	—	95x	370x	—	38x
1/16—5056—.001	9.5	1700p	—	1800p	—	500p	—	980p	—	110p	600p	—	50p
1/8—5056—.0007	**3.1**	**340**	**250**	**360**	**260**	**97**	**170**	**250**	**200**	**45.0**	**155**	**110**	**20.0**
1/8—5056—.001	**4.5**	**670**	**475**	**700**	**500**	**185**	**320**	**450**	**350**	**70.0**	**255**	**205**	**28.0**
1/8—5056—.0015	6.1	1180	760	1240	825	295	535	710	525	102	400	305	38.0
1/8—5056—.002	8.1	1800	1200	1900	1300	435	810	945	740	143	520	440	51.0
5/32—5056—.0007	2.6	260	180	275	185	70	120	200	152	36.0	120	80	17.0
5/32—5056—.001	3.8	530	360	560	375	140	235	360	272	57.0	205	155	24.0
5/32—5056—.0015	5.3	820	615	865	650	240	420	580	435	85.0	310	250	33.0
5/32—5056—.002	6.9	1220	920	1340	1000	350	650	760	610	118	430	360	43.0
3/16—5056—.0007	2.0	190	110	200	120	45	75	140	105	27.0	85	50	13.0
3/16—5052—.001	**3.1**	**405**	**250**	**420**	**260**	**97**	**170**	**270**	**200**	**45.0**	**155**	**110**	**20.0**
3/16—5056—.0015	4.4	660	460	690	490	180	310	425	340	68.0	245	198	27.5
3/16—5056—.002	5.7	960	685	1030	735	270	480	575	480	94.0	340	280	36.0
1/4—5056—.0007	1.6	100	75	110	80	30	50	90	78	20.0	60	38	12.0
1/4—5056—.001	**2.3**	**240**	**145**	**265**	**155**	**58**	**100**	**180**	**130**	**32.0**	**105**	**62**	**15.0**
1/4—5056—.0015	3.4	440	300	480	315	115	200	290	230	50.0	175	130	22.0
1/4—5056—.002	4.3	580	440	620	465	172	300	400	325	67.0	240	190	27.0
1/4—5056—.0025	5.2	790	600	820	645	230	410	500	425	84.0	300	245	32.0
3/8—5056—.0007	1.0	55	25	60	35	15	35	60	45	15.0	35	25	9.0
3/8—5056—.001	1.6	100	75	110	80	30	50	90	78	20.0	60	38	12.0
3/8—5056—.0015	2.3	205	155	210	155	58	100	170	130	32.0	105	62	15.0
3/8—5056—.002	3.0	320	240	340	260	92	160	245	190	43.0	145	100	19.0

Table 2-7. Properties of 2024 Hexagonal Aluminum Honeycomb (Ref. 2-2)

Hexcel Honeycomb Designation Cell — Material — Gage	Nominal Density pcf	Compressive					Crush Strength psi	Plate Shear					
		Bare		Stabilized				"L" Direction			"W" Direction		
		Strength psi		Strength psi		Mod ksi		Strength psi		Mod ksi	Strength psi		Mod ksi
		typ	min	typ	min	typ	typ	typ	min	typ	typ	min	typ
1/8—2024—.0015	5.0	810	525	840	620	200	425	500	400	82.0	315	250	33.0
1/8—2024—.002	6.7	1200	825	1330	980	300	640	760	600	118	470	375	45.0
1/8—2024—.0025	8.0	1750	1100	1900	1320	380	840	960	770	148	590	470	54.0
1/8—2024—.003*	9.5	2375	1475	2500	1725	480	1120	1150	950	170	650	585	64.0
3/16—2024—.0015	3.5	400	250	500	290	86	200	290	230	55.0	180	143	23.0
1/4—2024—.0015	2.8	290	165	320	175	40	110	200	140	42.0	120	88	19.0

*Not available with a normal CR III coating.

Notes:
Test data obtained at 0.625 inch thickness.
p = Preliminaries
x = Predicted values.

Table 2-8. Properties of ACG Hexagonal Aluminum Honeycomb (Ref. 2-2)

Hexcel Honeycomb Designation Material — Cell — Gage	Nominal Density pcf	Compressive			Crush Strength psi	Plate Shear			
		Bare	Stabilized			"L" Direction		"W" Direction	
		Strength psi	Strength psi	Modulus ksi		Strength psi	Modulus ksi	Strength psi	Modulus ksi
		typ	typ	typ	typ	typ	typ	typ	typ
ACG—1/4—.003	**5.2**	**620**	**650**	**14**	**245**	**345**	**63**	**215**	**31**
ACG—3/8—.003	**3.6**	**325**	**340**	**92**	**120**	**210**	**40**	**130**	**20**
ACG—1/2—.003	2.5	165	170	40	60	130	25	70	15
ACG—3/4—.003	**1.8**	**110**	**120**	**24**	**45**	**95**	**16**	**55**	**8**
ACG—1—.003	1.4	75	80	16p	25	55p	14p	40p	7p
ANG 2—3/8	4.2	600	620	135	220	300	65	190	29

Notes:
Test data obtained at 0.625 inch thickness.
p = Preliminaries
x = Predicted values.

Table 2-9. Properties of Corrugated Aluminum Honeycomb (Ref. 2-2)

Hexcel Honeycomb Designation Cell — Material — Gage	Nominal Density pcf	Crush Strength psi	Compressive Strength		Beam Shear Strength	
			Bare psi	Stabilize psi	"L" Direction psi	"W" Direction psi
CR-ALC—1/8—5052—.003	**12.0**	**1600**	**2300**	**2400**	**1950**	**1500**
CR-ALC—1/8—5052—.004	15.5	2350	3100	3200	2300p	1650p
CR-ALC—1/8—5052—.006	**22.1**	**4000**	**5100**	**5200**	**3000**	**2050**
CR-ALC—1/8—5052—.006 (R1)	27.5	4500	5800	5900	3700p	2100p
CR-ALC—1/8—5052—.006 (R2)	**38.0**	**6500**	**8500**	**8700**	**4300p**	**2200p**
CR-ALC—1/8—5052—.006 (2R2)	55.0	—	—	13000p	4900p	2610p
CR-ALC—3/16—5052—.003	**8.1**	**900p**	**1200p**	**1400p**	—	—
CR-ALC—3/16—5052—.004	10.6	1400	2200	2300	—	—
CR-ALC—3/16—5052—.006	**15.7**	**2400**	**3200**	**3300**	—	—
CR-ALC—3/16—5052—.006 (R2S)	25.0	4400p	5700	5800p	3350p	1700p
CR-ALC—1/4—5052—.006	**10.5**	**1350**	**2100**	**2200**	—	—

p = Preliminary values. Test data obtained at 0.625 inch thickness.
R1 = Reinforced every other ribbon.
R2 = Reinforced every ribbon.
2R2 = Double reinforced every ribbon.
R2S = Reinforced every ribbon, staggered.

Type II and ASTM E 1091 Type V and is frequently used in the manufacture of military rigid wall shelters. Other non MIL specification paper cores are used in construction curtain walls. Table 2-15 lists the requirements of ASTM E 1091 nonmetallic honeycomb cores. Type I requirements are for the aramid -fiber reinforced honeycomb core used in military shelters. The volatile content of the nonmetallic honeycomb core (including moisture) at the time of panel assembly shall not exceed 2 percent by weight when determined in accordance with ASTM E 874 Paragraph 10.4.2.

Solid Core Materials

Solid core materials are generally either made of balsa wood or of closed cell foams. These materials, like the honeycomb core, act as the web of an I beam in the distribution of loads through the structure.

***Balsa Wood*:** Ninety percent of all balsa wood is used as a structural core material. Balsa is very light but is also very strong. It is very soft but is actually a hardwood like oak. Under a microscope balsa is seen to be a high-aspect-ratio,

Table 2-10. Properties of Glass Reinforced Honeycomb (Ref. 2-2)

Hexcel Honeycomb Designation / Material - Cell - Density	Compressive					Plate Shear					
	Bare		Stabilized			"L" Direction			"W" Direction		
	Strength psi		Strength psi		Modulus ksi	Strength psi		Modulus ksi	Strength psi		Modulus ksi
Hexagonal	typ	min	typ	min	typ	typ	min	typ	typ	min	typ
HRP—3/16—4.0	**500**	**350**	**560**	**480**	**57**	**290**	**210**	**14**	**160**	**130**	**7**
HRP—3/16—5.5	**800**	**570**	**870**	**750**	**95**	**435**	**350**	**19**	**240**	**200**	**11**
HRP—3/16—7.0	1150	900	1230	1000	136	580	500	30	290	—	13
HRP—3/16—8.0	1440	1100	1530	1280	164	690	600	34	435	370	19
HRP—3/16—12.0	2400	1700	2520	1900	260p	930	815	48	620	525	28
HRP—1/4—3.5	**375**	**280**	**455**	**400**	**46**	**230**	**170**	**9**	**130**	**100**	**5**
HRP—1/4—4.5	**620**	**480**	**700**	**560**	**70**	**355**	**250**	**15**	**200**	**155**	**8**
HRP—1/4—5.0	700	510	820	660	84	390	—	20	240	—	10
HRP—1/4—6.5	1025	850	1180	900	120	500	—	25	260	—	13
HRP—3/8—2.2	165	120	200	145	13	115	90	6	65	45	2
HRP—3/8—3.2	**350**	**260**	**440**	**350**	**38**	**210**	**160**	**12**	**105**	**85**	**5**
HRP—3/8—4.5	610	450	690	550	65	300	260	14	190	150	8
HRP—3/8—6.0	900	750	1000	800	100	500	340	25	300	210	12
HRP—3/8—8.0	1060	920	1200p	—	150p	520p	—	31p	320	—	13p
OX-Core											
HRP/OX—1/4—4.5	580	460	650	515	43p	230	170	8.0	240	190	15
HRP/OX—1/4—5.5	775	580	890	—	65p	270p	—	10p	330	—	18
HRP/OX—1/4—7.0	1150	850	1230p	990p	84p	395p	—	14p	450p	—	20p
HRP/OX—3/8—3.2	340	260	425p	—	32p	140p	—	4.5p	150p	—	9p
HRP/OX—3/8—5.5	700	580	820p	—	60p	270p	—	10p	355p	—	17p
Flex-Core											
HRP/F35—2.5	180p	—	240p	—	25	125p	—	12p	70p	—	7.0p
HRP/F35—3.5	320p	—	400	300	37	200	140	15	105	75	10
HRP/F35—4.5	440p	—	600p	—	49	280p	—	22p	140p	—	12
HRP/F50—3.5	315	225	395	255	37	170	130	16	90	65	8
HRP/F50—4.5	420	340	600	500	49	265	200	25	140	100	13
HRP/F50—5.5	645	480	800	680	61p	420p	—	31p	235p	—	16p

Notes:
Test data obtained at 0.500 inch thickness.
p = Preliminary properties

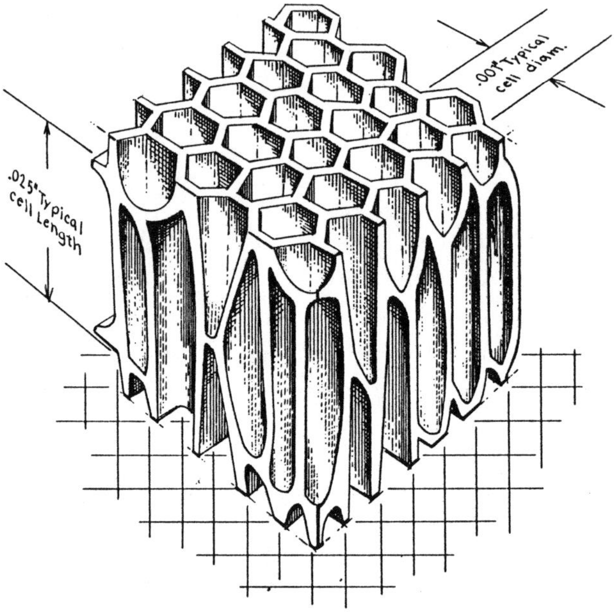

Figure 2-6. Drawing of the structure of end grain balsa. (Ref. 2-3)

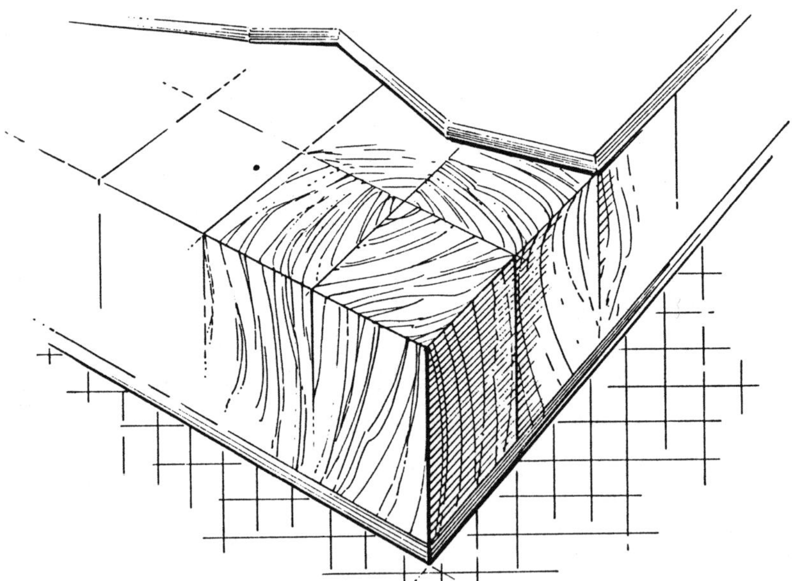

Figure 2-7. End-grain balsa in use as a sandwich core. (Ref. 2-3)

closed-celled structure. These high-aspect-ratio cells share large areas of wall structure between adjacent cells. Figure 2-6 is a magnified drawing showing the natural honeycomb structure of balsa. Under stress, this end-grain structure can carry enormous loads by distributing the loads from one cell to the

Table 2-11. Properties of Fibertruss Bias Weave Glass Reinforced Phenolic Honeycomb (Ref. 2-2)

Hexcel Honeycomb Designation Material - Cell - Density	Compressive			Plate Shear					
	Stabilized			"L" Direction			"W" Direction		
	Strength psi		Modulus ksi	Strength psi		Modulus ksi	Strength psi		Modulus ksi
	typ	min	typ	typ	min	typ	typ	min	typ
HFT—1/8—3.0	350	270	23	185	150	17	95	75	7.0
HFT—1/8—4.0	560	420	46	315	240	37	150	120	12
HFT—1/8—5.5	890	640	69	460	360	40	230	180	14
HFT—1/8—8.0	1750p	1500p	129p	580p	495p	49p	340p	290p	24p
HFT—3/16—1.8	120p	—	14p	105p	—	13p	50p	—	4.0p
HFT—3/16—2.0	170p	—	17p	115p	—	15p	60p	—	5.0p
HFT—3/16—3.0	365	275	34	200	155	24	100	80	9.0
HFT—3/16—4.0	590	430	44	275	210	30	140	115	14
HFT—3/16/OX—6.0	1180p	—	63p	320p	—	18p	260p	—	19p
HFT—3/8—4.0	500	350	—	300	210	24p	145	100	11p

Table 2-12. Properties of Glass Reinforced Polyester Honeycomb (Ref. 2-2)

Hexcel Honeycomb Designation Material - Cell - Density	Compressive					Plate Shear					
	Bare		Stabilized			"L" Direction			"W" Direction		
	Strength psi		Strength psi		Modulus ksi	Strength psi		Modulus ksi	Strength psi		Modulus ksi
Hexagonal	typ	min	typ	min	typ	typ	min	typ	typ	min	typ
NP—3/16—4.5	**450**	**355**	**600**	**500**	**80**	**280**	**195**	**14**	**130**	**90**	**5.2**
NP—3/16—6.0	850	615	1050	735	116	330	230	15	155	110	5.8
NP—3/16—9.0	1700	1375	1800	1260	180	520	370	20	280	200	7.5
NP—1/4—4.0	**435**	**305**	**560**	**390**	**68**	**235**	**205**	**13**	**120**	**85**	**6.0**
NP—1/4—6.0	940	650	1050	736	116	350	230	15	190	110	5.8
NP—1/4—8.0	1400	980	1540	1080	160	410	290	18	205	145	7.0
NP—3/8—2.5	180	140	265	195	34	170	120	10	100	70	4.0
NP—3/8—4.5	600	420	670	470	80	280	195	14	130	90	5.2
OX-Core											
NP/OX—1/4—4.0	**390p**	**340p**	—	—	—	**160p**	—	**5.0p**	**190p**	—	**12p**
NP/OX—1/4—6.0	700p	—	—	—	—	275p	—	7.5p	375p	—	19p
NP/OX—3/8—4.5	420p	—	—	—	—	190p	—	5.5p	285p	—	15p

Notes:
Test data obtained at 0.500 inch thickness.
p = Preliminary properties
HFT Glass Reinforced Phenolic Honeycomb is normally not tested for bare compressive strength.

Table 2-13. Properties of Glass Reinforced Polyimide Honeycomb (Ref. 2-2)

Hexcel Honeycomb Designation Material - Cell - Density	Compressive			Plate Shear					
	Stabilized			"L" Direction			"W" Direction		
	Strength, psi		Modulus, ksi	Strength, psi		Modulus, ksi	Strength, psi		Modulus, ksi
	typ	min	typ	typ	min	typ	typ	min	typ
HRH-327—1/8—3.2	310	220	2.7	195	140	19	95	70	7.5
HRH-327—1/8—5.5	790p	600p	—	465p	300p	30p	245p	175p	14.5p
HRH-327—3/16—4.0	440p	—	50p	280p	—	29p	130p	—	10p
HRH-327—3/16—4.5	**520**	**400**	**58**	**320**	**220**	**33**	**150**	**110**	**11**
HRH-327—3/16—5.0	600p	—	68p	370p	—	37p	180p	—	12p
HRH-327—3/16—6.0	780	625	87	460	345	45	230	170	15
HRH-327—3/16—8.0	**1210**	**1000**	**126**	**700**	**490**	**55**	**420**	**300**	**22**
HRH-327—3/8—4.0	**440**	**325**	**50**	**280**	**195**	**29**	**150**	**100**	**12**
HRH-327—3/8—5.5	680	540	78	420	300	41	210	160	13
HRH-327—3/8—7.0	1000p	—	106p	550p	—	53p	310p	—	18p

next. Balsa sandwich structures have high impact and fatigue resistance; excellent flexural, shear and bond strengths; and exceptional resistance to high temperature and fire. Balsa also provides superior sound and thermal insulation and because it is closed-cell (less than 10 percent of balsa's end-grain structure consists of open vessels) and resists water migration across the grain. Figure 2-7 depicts end-grain balsa being used as a core in a sandwich structure. Balsa core has many uses in marine, transportation, industrial, architectural and military applications. Some examples of the use of balsa core include racing and cruising boats, commercial fishing vessels, aircraft flooring, storage units, partitions, galleys, rail and highway container bodies, wind turbine blades, decorative panel portable shelters and air cargo and air drop pallets. Some of the properties of end-grain balsa can be found in Table 2-16.

Table 2-14. Properties of Aramid -fiber Reinforced Honeycomb (Ref. 2-2)

Hexcel Honeycomb Designation / Material - Cell - Density - Gage	Compressive: Bare Strength psi		Compressive: Stabilized Strength psi		Compressive: Stabilized Modulus ksi	Plate Shear: "L" Direction Strength psi		Plate Shear: "L" Direction Modulus ksi	Plate Shear: "W" Direction Strength psi		Plate Shear: "W" Direction Modulus ksi
Hexagonal	typ	min	typ	min	typ	typ	min	typ	typ	min	typ
HRH-10—1/16—3.4 (1.5)	195p	150p	205p	160p	20	155p	125p	6.0p	85p	65p	2.9p
HRH-10—1/8—1.8 (1.5)	110	70	130	85	—	90	65	3.7	50	36	2.0
HRH-10—1/8—3.0 (2)	300	210	350	245	20	175	150	6.5	97	75	3.5
HRH-10—1/8—4.0 (3)	450	330	500	370	28	280	225	10.7	135	110	4.6
HRH-10—1/8—5.0 (3)	740	600	810	660	—	325	235	—	175	120	—
HRH-10—1/8—6.0 (3)	1035	800	1125	870	60	395	290	13.0	200	145	6.5
HRH-10—1/8—8.0 (3)	1605	1200	1780	1250	78	470	355	16.0	250	210	10.0
HRH-10—1/8—9.0 (3)	2170	1550	2300	1690	90	515	405	20.0	300	220	11.5
HRH-10—5/32—5.0 (4)	650p	520p	700p	550p	—	360p	—	11.0p	180p	—	5.0p
HRH-10—5/32—9.0 (4)	1725p	1340p	1900p	1490p	—	525p	—	18.0p	285p	—	9.5p
HRH-10—3/16—1.8 (2)	120	85	135p	—	—	90	70	3.8	50	40	1.9
HRH-10—3/16—2.0 (2)	155	105	165	115	11	110	80	4.5	60	45	2.7
HRH-10—3/16—3.0 (2)	300	210	370	270	20	160	120	5.8	88	68	3.5
HRH-10—3/16—4.0 (3)	500	360	560	425	28	245	200	8.6	135	100	5.0
HRH-10—3/16—4.5 (5)	425	320	475	400	—	290	225	9.5	145	110	4.0
HRH-10—3/16—6.0 (5)	940	660	1035	725	—	500	355	14.5	265	195	6.5
HRH-10—1/4—1.5 (2)	90	45	95	55	6	75	45	3.0	35	23	1.5
HRH-10—1/4—2.0 (2)	155	100	170	110	11	115	81	4.2	60	43	2.2
HRH-10—1/4—3.1 (5)	265	185	300	240	—	160	135	7.0	85	60	3.0
HRH-10—1/4—4.0 (5)	450	310	500	360	—	240	200	8.5	140	105	4.2
HRH-10—3/8—1.5 (2)	90	45	95	55	6	75	45	3.0	35	23	1.5
HRH-10—3/8—2.0 (2)	150	80	170	105	11	110	72	4.2	55	36	2.2
HRH-10—3/8—3.0 (5)	285p	—	300p	—	17p	170p	—	5.6p	95p	—	3.0p

Notes:
Test data obtained at 0.500 inch thickness.
p = Preliminary properties

Table 2-15. ASTM E1091 Requirements for Nonmetallic Honeycomb Core

Type	Type I	Type II	Type III	Type IV	Type V
Maximum density, lb/ft^3 (kg/m^3)	5.5 (88)	4.4 (70)	3.3 (53)	4.4 (70)	3.3 (53)
pH, pH units	7.0 ± 0.5	7.0 ± 0.5	7.0 ± 0.5	6.0 ± 0.5	6.0 ± 0.5
Compressive strength,^{A} psi (MPa):					
Dry, min^B	464 (3.20)	464 (3.20)	232 (1.60)	404 (2.79)	198 (1.37)
Wet, min^C	406 (2.80)	232 (1.60)	116 (0.80)	163 (1.13)	98 (0.68)
At elevated temperature^D	394 (2.72)	278 (1.92)	139 (0.96)	185 (1.28)	115 (0.79)
Cyclic aging, minimum	363 (2.50)	190 (1.31)	104 (0.72)	119 (0.82)	62 (0.43)
After fungus test^E			See *footnote*^{E}		
Shear strength,^{A} psi (MPa):					
Dry, min					
TL core orientation^B	218 (1.50)	218 (1.50)	116 (0.80)	180 (1.24)	85 (0.59)
TW core orientation^B	130 (0.90)	116 (0.80)	65 (0.45)	113 (0.78)	57 (0.39)
Wet, min					
TL core orientation^C	203 (1.40)	109 (0.75)	58 (0.40)	86 (0.59)	53 (0.37)
TW core orientation^C	116 (0.80)	58 (0.40)	33 (0.23)	58 (0.40)	32 (0.22)
Flatwise tensile strength, min,^{F}:					
psi (MPa)	406 (2.80)	406 (2.80)	231 (1.60)	306 (2.11)	231 (1.60)
Brittleness/Impact:					
Drop height, min, in. (mm)	40 (1016)	30 (762)	20 (508)	30 (762)	20 (508)
Water migration resistance:					
24 h, max, number of cells	3	3	3	3	3

^{A} Two-inch (51-mm) thick core with 0.05-in. (1.3-mm) facings, tested at 73 ± 2°F (22.7 ± 1°C) unless otherwise stated herein. Shear tests shall be conducted with the core oriented in the TL and TW planes (Fig. 1).
^{B} At equilibrium with 73 ± 2°F (22.7 ± 1°C), and 50 ± 4 % RH.
^{C} After soaking in water at 70 ± 5°C (21.1 ± 3°C) for 48 h with perforated facings.
^{D} After heating for 30 min at, and tested at, 176 ± 5°F (80 ± 3°C).
^{E} Values for compressive strength shall be not less than 85 % of the dry compressive strength values for Types I, II, and III and 80% for Types IV and V.
^{F} Tested at 73 ± 2°F (22.7 ± 1°C) with loading blocks bonded directly to each side of core specimen having a minimum area of 9 in.^{2} (5806 mm^2).

Table 2-16. Properties of End-grain Balsa (Ref. 2-3)

PROPERTY	UNITS	VALUE					
Density	Pounds/ft^3 (kg/m^3)	6 (96)	7 (110)	8 (130)	9½ (150)	11 (180)	15½ (250)
Thermal Conductivity ASTM C-177 @ 75°F	Btu-in/hr-ft^2-°F (W/m-°K)	0.353 (0.0509)	0.380 (0.0548)	0.408 (0.0588)	0.450 (0.0649)	0.492 (0.0710)	0.617 (0.0890)
Tensile Strength ASTM C-297	psi (MPa)	1,000 (6.90)	1,320 (9.10)	1,550 (10.7)	1,900 (13.1)	2,280 (15.7)	3,440 (23.7)
Shear Strength ASTM C-273	psi (MPa)	268 (1.85)	314 (2.17)	361 (2.49)	432 (2.98)	502 (3.46)	717 (4.94)
Shear Modulus ASTM C-273	psi (MPa)	15,600 (108)	17,400 (120)	19,500 (134)	23,100 (159)	27,300 (188)	45,300 (312)
Compression Strength ASTM C-365	psi (MPa)	945 (6.52)	1190 (8.21)	1450 (10.0)	1870 (12.9)	2320 (16.0)	3850 (26.6)
Compression Modulus ASTM C-365	psi (MPa)	325,000 (2,240)	397,000 (2,740)	472,000 (3,260)	590,000 (4,070)	715,000 (4,930)	1,120,000 (7,720)
Linear Coefficient of Thermal Expansion	(IN/IN/°F)	Tangential Radial Longitudinal	10.5×10^{6} 7.0×10^{6} 1.7×10^{6}				

Foam Core: The foam materials most frequently used in sandwich construction are closed cell polyurethane foams. Polystyrene foams and other foam materials such as the low density syntactic foams are taking their place in the structural materials. The properties of foam materials can vary depending on the formulation, application and processing. The materials used for military shelter repair must meet one of the following specifications: (1) MIL-P-26514 Polyurethane Foam, Class 1-rigid, nominal density 4.0 pounds per cubic foot; or (2) MIL-C-83400 Type II Core Material for Metal Sandwich Panels for Shelter Construction (polyurethane foam), nominal density 4.0 pounds per cubic foot.

Foam material may be fabricated in large blocks and then cut into slabs or panels which are then bonded between skins to form the structure, or they can be foamed in place using a liquid foam material. The foamed in place technique is use to produce structural architectural panels in which the liquid foaming material is spread between two skins and allowed to expand under controlled restraints. Liquid foam material may also be used for the repair of panels instead of solid core material as in the repair of military shelters.

The newer syntactic core materials are prepared, shipped and handled in the same manner as prepreg materials. The core material is cured under controlled conditions of temperature and pressure. Foam core material is also used as sheets of frangible foam forced under compression into the cells of nonmetallic honeycomb core to increase the thermal insulation of the panel.

REFERENCES

2-1 "The Basis on Bonded Sandwich Construction" Hexcel TSB 124 (1986)

2-2 "Mechanical Properties of Hexcel Honeycomb Materials" Hexcel TSB 120 (1988)

2-3 "Why Balteck Core" Balteck Corporation Data File 159

3

DAMAGE ASSESSMENT AND REPAIR METHOD SELECTION

GENERAL

Before a structure is committed for repair it should be submitted for damage assessment. This involves a series of inspections and evaluations to determine if repair is justified. A careful assessment of the extent of the damage allows for the definition of the damage, the amount of the structure which must be repaired or replaced, and is the basis of an accurate estimate of the costs involved. But foremost the assessment is the basis for the safety evaluations required before the structure can be returned to service.

Early detection of damage is important. Holes in a structure must be attended to as soon as they are detected. This should be done on site. It is important to keep moisture and other foreign materials out of the damaged area. There are three classifications of repair defined as follows:

Temporary Preservation - The prevention of the entrance of foreign material in to the defect or damaged area until the damage can be repaired. This should be done as soon as the damage is discovered.

Semi-permanent Repair - Restoration of a structure to full serviceability but requiring increased inspection frequencies to ensure that the repair remains effective. This type of repair is often made at a field location, and will generally be removed at the next major overhaul.

Permanent Repair - Restoration of the structure to its original integrity.

DAMAGE ASSESSMENT

Methods used for damage assessment are discussed under Nondestructive Inspection. Typical damage assessment includes the following:

1. **External Damage** - This is a visual inspection of the structure looking for obvious damage such as punctures and indications of delamination or impact. All punctures should be covered immediately to prevent further damage (temporary preservation).
2. **Delamination** - Delamination of a skin-to-core bond and the top layers of a laminate can often be detected by visual inspection along with tapping of the surface. Detection of non-surface delaminations requires some form of nondestructive testing.
3. **Impact Damage** - Impact damage in a composite material is generally difficult to detect since the real danger is not seen on the surface but is propagated to the other side of the laminate. This type of damage can be detected nondestructively by dry contact acoustic transmission, an acousto-ultrasonic technique. (Ref. 3.1.)
4. **Internal Defects** - Internal defects such as voids, distorted core or crushed core can only be detected by nondestructive techniques.
5. **Bondline Degradation** - Evaluation of adhesive degradation can be made nondestructively by the dry contact acoustic transmission technique.
6. **Skin Cracks** - Skin cracks and laminate crazing can often be determined by visual inspection.

7. **Fuel and Oil Contamination** - The structure should be inspected for signs of fuel and oil contamination. If fuel has penetrated the structure, both safety and structural integrity must be considered before a repair is attempted.
8. **Moisture** - Moisture in a structure can be determined by X-ray and by some thermal techniques.
9. **Corrosion** - Inspection for corrosion generally involves examining the structure for signs of corrosion products, swelling of the metal surface, etc. If corrosion damage has occurred, it is necessary to remove all traces of the corrosion before repair is made. If any corrosion product is left in the structure it will grow and cause further damage.

A description of the various damages modes apply to the assessment of damage in structures that are to be repaired are given in Table 3-1.

REPAIR METHOD SELECTION

A guide to the appropriate method of repair for sandwich and composite laminate structures is found in Table 3-2. The use of these methods is subject to limitations stipulated in the specific items repair manual. Extension of these methods to the large area repairs must be authorized by the responsible engineering agency. In general, emphasis is placed on using methods that will minimize subsequent problems. The approach is to use properly prepared surfaces, impermeable heat and pressure cured adhesives, and adequately sealed panels or patch edges. The latter involves the use of a minimum width peripheral bondline plus a sealant. The patch-type approach is generally most expedient and economical when it can be used. For larger parts or highly stressed areas, it may be necessary to replace entire detail parts. Replacement of damaged details allows the component to be restored to its original strength and contour.

Rebuilding or replacement of detailed parts, e.g., skins, may be in order when high service temperatures are encountered. This is because of the lower strength of adhesives at higher temperatures, hence their limitation in transferring structural loads. If a patch repair is used in these areas it may be necessary to classify the repair as a temporary repair and require more frequent inspections.

Table 3-1 Description of Damage Modes

Damage Mode	Alternate Description	Definition
Break		Gap or opening in a structure.
Crack	Fracture	A narrow break or fracture in the surface or through the material.
Crushed	Collapsed	Deformation caused by a compressive overload (usually transverse) which causes the material to be permanently displaced from its natural shape or form.
Debond	Disbond	Separation of an adhesive bondline.
Dent		An indentation or depression which does not rupture the material.
Gouge	Groove	A groove, channel, or cavity scooped or cut from the surface of the material.
Hole	Puncture Penetration	An opening or cavity in or through the material.
Tear	Split	A jagged, irregular split or rupture in the material.

Table 3-2. Guide To Repair Method Selection

Type of Damage	Type of Structure	Repair Method
Minor	Laminate, nonstructural	6.1
Major	Laminate, nonstructural	6.2
Major	Laminate, structural	6.3
Major	Laminate, structural	6.4
Major	Laminate, structural	6.5
Dents	Honeycomb, composite skin	6.6
Tears, Cracks and surface defects	Honeycomb, composite skin	6.7
Minor core	Honeycomb, composite skin	6.8
Major, one skin	Honeycomb, composite skin	6.9
Major, both skins	Honeycomb, composite skin	6.10
Edge voids	Honeycomb, composite skin	6.11
Large area, one skin	Honeycomb, composite skin	7.1
Large area, both skins	Honeycomb, composite skin	7.2
Rebuild	Honeycomb, composite skin	7.3
Dents	Honeycomb, metal skin	8.1
Dents with void	Honeycomb, metal skin	8.2
Dent with void	Honeycomb, metal skin	8.3
Dents with crack	Honeycomb, metal skin	8.4
Minor core	Honeycomb, metal skin	8.5
Major, one skin	Honeycomb, metal skin	8.6
Major, both skins	Honeycomb, metal skin	8.7
Edge voids	Honeycomb, metal skin	8.8
Large area, one skin	Honeycomb, metal skin	9.1
Large area, both skins	Honeycomb, metal skin	9.2
Rebuild	Honeycomb, metal skin	9.3

Puncture	Foam, metal skin	10.1
Through hole	Foam, metal skin	10.2
Minor,	Foam, metal skin	10.3
Major, one skin	Foam, metal skin	10.4
Extensive	Foam, metal skin	10.5
Spanning internal member	Foam, metal skin	10.6
Damaged structural member	Foam, metal skin	10.7
Puncture	Balsa, metal skin	10.8
Through hole	Balsa, metal skin	10.9
Major, both skins	Balsa, composite skin	10.10

When the same material is not available for the fabrication of metal parts the information given in Table 3-3 can be used to determine the selection of replacement materials. (Ref. 3-1.) Table 3-4 shows the minimum acceptable rivet edge distances for aluminum alloy rivets (both cherry and lockbulb types). (Ref. 3.2.)

REFERENCES

3-1 Department of the Army Technical Manual TM-55-1520-228-bd. DS & GS Maintenance Manual Army Model OH58A Helicopter Nov 1972

3-2 Department of the Army Technical Manual TM 55-1500-204-25/1 General Aircraft Maintenance Manual April 1970

Table 3-3. Metal Substitution Chart

MATERIAL TO BE REPLACED	ULTIMATE TENSILE STRENGTH PSI	2024			1025	7075 T6		4130 8630 STEEL		TITANIUM				STAINLESS STEEL		
		T3&4 CLAD	T4 EXTRUDED	T6 BARE	STEEL	CLAD	EXTRUDED	90 KSI	125 KSI	99%	8 Mn	6AL -4V	6Al-6V-2Sn	1/4 301	1/2 302	321 347
6061-T6 EXTRUDED	38,000	1.0	1.0	1.0	1.0	1.0	1.0	1.0	1.0	1.0	1.0	1.0	1.0	1.0	1.0	1.0
AZ31A-H MAGNESIUM	39,000	1.0	1.0	1.0	1.0	1.0	1.0	1.0	1.0	1.0	1.0	1.0	1.0	1.0	1.0	1.0
6061-T6 CLAD	42,000	1.0	1.0	1.0	1.0	1.0	1.0	1.0	1.0	1.0	1.0	1.0	1.0	1.0	1.0	1.0
2024-T4 EXTRUDED	57,000	1.0	1.0	1.0	1.04	1.0	1.0	1.0	1.0	1.0	1.0	1.0	1.0	1.0	1.0	1.0
2024-T4 CLAD	58,000	1.0	1.02	1.0	1.05	1.0	1.0	1.0	1.0	1.0	1.0	1.0	1.0	1.0	1.0	1.0
2024-T3 CLAD	60,000	1.04	1.05	1.0	1.09	1.0	1.0	1.0	1.0	1.0	1.0	1.0	1.0	1.0	1.0	1.0
2024-T6 BARE	62,000	1.07	1.09	1.0	1.13	1.0	1.0	1.0	1.0	1.0	1.0	1.0	1.0	1.0	1.0	1.0
2024-T81 CLAD	64,000	1.1	1.12	1.03	1.16	1.0	1.0	1.0	1.0	1.0	1.0	1.0	1.0	1.0	1.0	1.0
2024-T88 CLAD	70,000	1.21	1.23	1.13	1.25	1.0	1.0	1.0	1.0	1.0	1.0	1.0	1.0	1.0	1.0	1.0
7075-T6 CLAD	72,000	1.24	1.27	1.16	1.31	1.0	1.0	1.0	1.0	1.0	1.0	1.0	1.0	1.0	1.0	1.0
7075-T6 BARE	78,000	1.35	1.37	1.26	1.42	1.09	1.0	1.0	1.0	1.0	1.0	1.0	1.0	1.0	1.0	1.0
7075-T6 EXTRUDED	78,000	1.35	1.37	1.26	1.42	1.09	1.0	1.0	1.0	1.0	1.0	1.0	1.0	1.0	1.0	1.0
7178-T6 BARE	84,000	1.45	1.48	1.36	1.53	1.17	1.08	1.0	1.0	1.0	1.0	1.0	1.0	1.0	1.0	1.0
7178-T6 EXTRUDED	84,000	1.45	1.48	1.36	1.53	1.17	1.08	1.0	1.0	1.0	1.0	1.0	1.0	1.0	1.0	1.0
TITANIUM 99Z	80,000	1.38	1.40	1.29	1.45	1.11	1.02	1.0	1.0	1.0	1.0	1.0	1.0	1.0	1.0	1.0
TYPE 321 & 347 CRES	100,000	1.72	1.75	1.61	1.82	1.38	1.28	1.11	1.0	1.25	1.0	1.0	1.0	1.0	1.0	1.0
TITANIUM 8Mn	120,000	2.06	2.1	1.93	2.18	1.66	1.53	1.33	1.0	1.50	1.0	1.0	1.0	1.0	1.0	1.20
TYPE 301 STAINLESS	125,000	2.15	2.19	2.09	2.27	1.73	1.60	1.38	1.0	1.56	1.04	1.0	1.0	1.0	1.0	1.25
TITANIUM 6AL-4V	134,000	2.31	2.35	2.16	2.43	1.86	1.71	1.48	1.07	1.67	1.12	1.0	1.0	1.07	1.0	1.34
TITANIUM 4AL-4Mn	140,000	2.41	2.45	2.25	2.55	1.94	1.79	1.55	1.12	1.75	1.16	1.04	1.00	1.12	1.0	1.40
TYPE 301 STAINLESS	150,000	2.58	2.63	2.42	2.73	2.08	1.92	1.66	1.20	1.88	1.25	1.11	1.0	1.2	1.0	1.50
TITANIUM 6Al-6V-2Sn	155,000	2.67	2.7[illegible]	2.5	2.81	2.15	1.98	1.72	1.24	1.93	1.29	1.15	1.0	1.24	1.03	1.55

A. LOCATE THE MATERIAL TO BE REPLACED ON THE LINE IN THE LEFT HAND COLUMN.
B. LOCATE THE SUBSTITUTE MATERIAL IN THE VERTICAL COLUMNS.
C. TO OBTAIN THE MINIMUM THICKNESS OF THE SUBSTITUTE MATERIAL, MULTIPLY THE THICKNESS OF THE MATERIAL TO BE REPLACED BY THE FACTOR SHOWN AT THE INTERSECTION OF THE LINE AND COLUMN FOUND IN STEP A & B. SUBSTITUTE STANDARD GAGE EQUAL TO THIS THICKNESS OR NEAREST STANDARD GAGE.

Table 3-4. Minimum Acceptable Rivet Edge Distances For Aluminum Alloy Rivets.

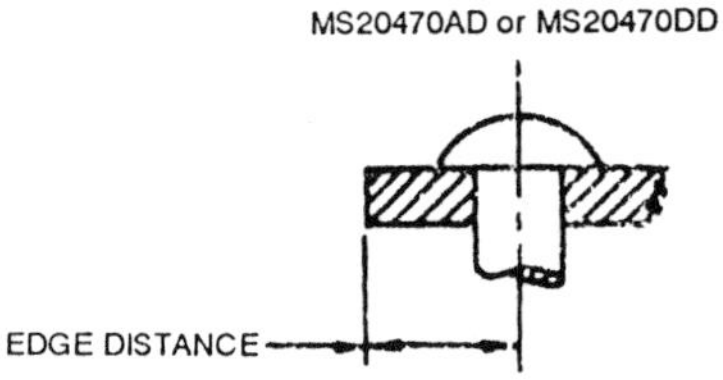

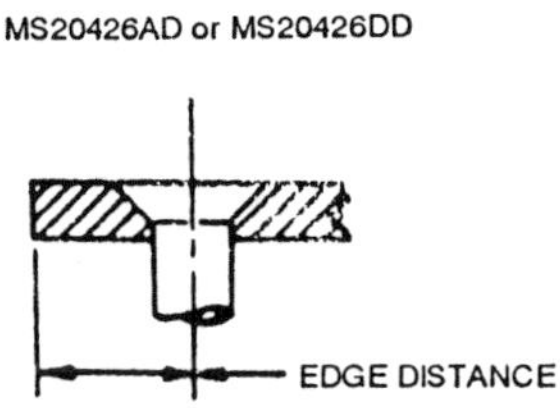

MINIMUM RIVET EDGE DISTANCES

SHEET THICKNESS \ RIVET DIA IN 32NDS	NONFLUSH-TYPE RIVETS				FLUSH-TYPE RIVETS			
	3	4	5	6	3	4	5	6
0.020	0.156	0.218	0.265	--	0.187	0.281	0.328	--
0.025	0.156	0.218	0.265	0.328	0.187	0.281	0.328	0.390
0.032	0.140	0.218	0.265	0.328	0.171	0.281	0.328	0.390
0.040	0.125	0.187	0.234	0.328	0.156	0.250	0.328	0.390
0.051	0.109	0.156	0.203	0.312	0.140	0.218	0.296	0.375
•0.064	--	0.140	0.187	0.265	--	0.203	0.265	0.328
•0.072	--	0.140	0.171	0.250	--	0.203	0.250	0.312
•0.081	--	0.125	--	0.234	--	--	0.234	0.296
•0.102	--	--	--	0.203	--	--	--	0.265

• USE MS20470DD OR MS20426DD RIVETS

NOTES

1. ALL DIMENSIONS ARE SHOWN IN INCHES

2. MINIMUM EDGE DISTANCES SHOWN ARE APPLICABLE ONLY TO ROWS OF RIVETS CONTAINING THREE OR MORE RIVETS, AND ARE FURTHER LIMITED TO A MAXIMUM OF 30 PERCENT OF TOTAL RIVETS IN A ROW

4

SURFACE PREPARATION PROCEDURES

Adhesion is controlled by the condition of the surface of the adherends and the environment in which the bond must function. The following is a guideline for surface preparation of adherends:

1. Know the adherend material, how it was fabricated and what is its surface condition.
2. Use good safety and health practices when using cleaning agents.
3. For wiping parts, only clean white cotton cloths with no sizing or finish or paper towel with little or no organic binders should be used. Never use synthetic cloths.
4. Hot solvent washing or vapor degreasing should not be used to clean composite materials or for secondary bonding preparations.
5. Check quality control (QC) of the cleaning baths regularly.
6. Train, qualify and requalify personnel who operate the cleaning facility.
7. Cleanliness, vigilance and attention to details of the process are of primary importance.

8. Don't rely on faith. Know the process, the materials, personnel and quality control status at all times.
9. Follow the process. Don't change the temperatures, times, etc, unless authorized by the cognizant process engineer.

PREPARATION OF COMPOSITE SURFACES

Preparation of composite faying surfaces is a critical step in the adhesive bonding process. These surfaces, by their nature, can not be prepared by the conventional chemical type treatments. The faying surface of a composite is prepared by mechanical means such as machining, sanding or grinding. The care and dedication with which the preparation of a composite surface is carried out will largely determine the success or failure of the bonded repair. Composites containing fiberglass and graphite can usually be prepared with standard tools and abrasive materials. However, some special tools will result in longer tool life. Composites containing Kevlar often require special tools and operating conditions for successful machining. Special tools are listed in Table 2-1. Suggested operating conditions for machining composites containing Kevlar are given in Table 2-3. A mechanical scarfing machine reduces time and improves scarf angle accuracy (Ref. 4-1) The scarfing tool is discussed further in Chapter 5, "Tools, Equipment and Facilities."

PREPARATION OF COMPOSITE SURFACES FOR BONDING

Hand sanding of composite surfaces can be accomplished as follows:

1. Remove the paint with 180 grit sand paper or equivalent.
2. Wipe the surface with safety solvent.
3. Lightly abrade the surface with 180 grit (or other specified grit) abrasive paper or equivalent.
4. Remove debris with a vacuum.

5. Protect the surface with wax-free paper until the assembly is to be bonded.

PREPARATION OF METALLIC SURFACES FOR BONDING

Preparation of Aluminum - Phosphoric Acid Nontank Anodize Process (Phos-Anodize Containment System, PACS)

Preparation of the Surface to be Anodized:

1. Remove all dirt, oil and contaminants from the surface of the repair area by solvent cleaning.
2. Remove any paint from the repair area with paint stripper and/or mechanical abrasion. **CAUTION: Do not allow any stripper to enter the interior of the bonded structure.**
3. Clean the repair area using safety solvent.
4. Lightly abrade the surface with Scotch-Brite, Type A very fine pad or equivalent and wipe with safety solvent to obtain a water-break-free surface. NOTE: A water-break-free surface is one which maintains a continuous water film (i.e., no beads) for a period of at least 30 seconds after having been spray- or immersion-rinsed in clean water at a temperature below 100 degrees F.
5. Verify the water-break-free surface for 30 seconds. If the film does not remain continuous for 30 seconds, repeat Step 3.
6. Dry the surface with clean gauze, and anodize the prepared surface immediately.

Preparation and assembly of the PACS: The PACS must be assembled correctly to provide the conditions for anodizing as well as to safely contain the phosphoric acid solution.

1. Mask cracks, honeycomb and all areas not to be anodized. **CAUTION: Do not allow phosphoric acid to contaminate fasteners, crevices, and undamaged areas. Mask off these**

areas with protective anodizing tape. Do not allow the acid to enter the honeycomb core.

2. Remove the PACS kit from the shipping box and separate the acid-rinse solutions from the kit and the waste collection bladder.
3. Remove the backing paper from the inner vacuum bag sealant.
4. Place the inner vacuum bag over the repair area and draw a vacuum using the ATACS 8024 Hot-Bonder. See Figure 4-1.
5. Remove the backing paper from the outer vacuum bag sealant and place over the inner bag. Draw a vacuum on the outer bag.
6. Connect the tubing for the acid and rinse bladders to the three way valve.
7. Connect the waste collection bladder.

Anodizing:

1. Connect the PACS power supply cord to the ATACS 8024 Hot-Bonder's power source (the 8010 PACS kit plug will only fit the correct connection on the 8024 Hot-Bonder, thus eliminating the possibility of incorrect power).
2. Turn the three-way valve to the acid position.
3. Turn on the power. NOTE: The PACS kit incorporates an orifice designed to allow the correct amount of acid into the inner bag and on to the repair area. The waste containment bladder will collect all the used acid and rinse solution.
4. When all the acid has been drawn through the bag (approximately 20 minutes) turn the three way valve to allow the area to rinse. Turn off the power.

Disassembly of PACS: Removal of the PACS assembly requires protection for personnel as well as special handling of the anodized surface. **WARNING: Do not allow acid to contact**

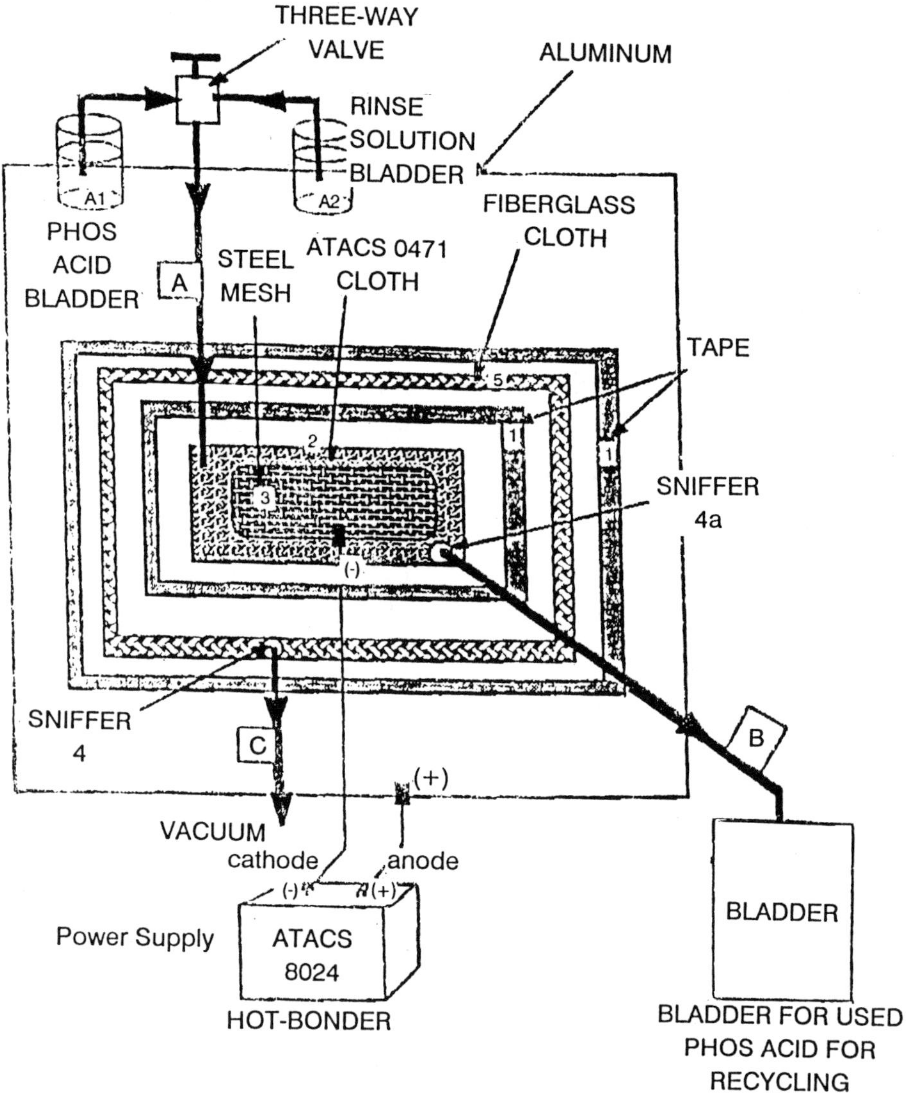

Figure 4-1: Schematic for the Atacs 8010 phos-anodize containment system.

skin or eyes. Wear rubber gloves, eye and face protection, and protective clothing. If contact does occur, immediately flush affected area with large quantities of water and seek medical aid. Acid will burn skin and eyes if contact is made.

1. Turn off and remove all vacuum sources and disconnect all tubing.
2. Remove the outer bag with extreme care since it may contain be acid.
3. Remove the inner bag and the bleeder/screen layup, again being careful with any residual liquids.
4. Dispose of the consumable materials in an approved container adhering to approved methods of disposal.
5. Additional immersion or spray rinsing should be used if possible to remove any residual acid from the part.
6. Lightly wipe the area with a gauze moistened with water.
7. Rinse immediately (within 2.5 minutes) for a minimum of 5 minutes.
8. Air dry for a minimum of 30 minutes at room temperature or force air dry at 140-160 degrees F. (60-71 degrees C.) for 15 minutes. **CAUTION: Do not touch the dried anodized surface. Do not apply tape to the surface. Protect the anodized surface from contamination. Wear clean white cotton gloves when handling parts.**
9. Check the quality of the prepared surface. A properly anodized surface will show an interference color when viewed through a polarized filter rotated 90 degrees at a low angle of incidence to fluorescent light or daylight.
10. Prime the surface within one hour maximum.

Preparation of Aluminum - P2 Etch Tank Process

Materials:

1. *Water:* Deionized water must be used for processing

solutions and for rinses. Rinses should be maintained to prevent carryover of materials that would adversely effect the next solution.

2. *Degreasing:* Oils and greases are to be removed when necessary, by vapor degreasing, solvent cleaning, emulsion cleaning or alkaline cleaning. Solvent cleaning should be accomplished using safety solvent (O-T-620) at room temperature. Vapor degreasing may be accomplished with any of the following solvents, dependent upon local regulations; 1,1,1-trichloroethane, trichloroethylene, or perchloroethylene.

3. *Alkaline Cleaning Solution:* Non-etching alkaline cleaning solutions are to be prepared and used in accordance with the manufacturer's instructions. When the aluminum being cleaned is immersed in the alkaline cleaner for the time and at the temperature specified for processing, there should be no evidence of gas evolution, etching or metal removal. The alkaline cleaner must not contain silicates. Alkaline cleaning may be repeated as required. Keep the parts wet between the alkaline treatment and immersion in the rinse tank to prevent staining.

4. *P2 Etch Solution:* The etch solution should be maintained, as determined by chemical analysis, at approximately 27 to 36% by weight of sulfuric acid and 2.9 to 4.7 oz/gal of ferric iron.

5. *P2 Etching:* Immerse the aluminum parts in the etch solution for 10 to 12 minutes at a temperature maintained at 140 to 149 degrees F. (60 to 65 degrees C.). Keep the parts wet between the etch tank and the rinse tank by use of a spray rinse. Follow the etching by a thorough rinsing in room temperature deionized water.

   NOTE: Parts that are to be rerun because of lack of water-break-free surface, stains, or unprimed parts that have exceeded the permitted storage time, may be reworked no more than twice. Do not exceed a total immersion time of 34 minutes.

6. *Final Rinse:* Rinse the aluminum for one to two minutes in deionized water at a temperature between room temperature and 122 degrees F. (50 degrees C.).
7. *Drying:* Air dry the parts for not more than one hour at a temperature not to exceed 149 degrees F. (65 degrees C.).
8. *Prime:* As required.

PREPARATION OF OTHER METALS

Information on processing other metal surface may be found in "Surface Preparation Techniques for Adhesive Bonding" (Ref. 4-2) and in MIL-HDBK-337. (Ref. 4-3)

PAINT REMOVAL

Paint Removal from Metal Parts by Solvents

Paint may be removed from metal surfaces by the use of methyl-ethyl-ketone (MEK) solvent. **CAUTION: MEK usage may be restricted in some areas as it is a health hazard if proper safety precautions are not adopted. Check local safety regulations.**

Paint Removal by Mechanical Sanding

Metal Skins: The paint may be removed from metal skin structures by sanding with 80 or 100 grit sand paper followed by a light sanding with 400 grit sand paper.

Composite Skins: The paint may be removed from composite faced structures by a light sanding with 400 grit sand paper. Care is required so as to not sand down into the fiber of the composite material.

Paint Removal by Plastic Media Blasting (PMB)

The paint may be removed from metal and the following

composite surfaces: fiberglass, and carbon or graphite fiber reinforced materials. The method may not be used on engineering plastics whose design requirements specify that they remain transparent or translucent nor may it be used on Kevlar materials.

Plastic Media: The plastic media used to remove paint shall be a Type V Acrylic (Thermoplastic), Grade A which has been manufactured from virgin plastics and have a color No. 5, light gray. The material shall conform to MIL-P-85891. The particle shape of the acrylic media shall be irregular with sharp, angular edges and corners and should have a hardness of 46 to 54 BARCOL. The particle size should be a nominal 20 to 30 mesh.

Equipment Required: The PMB equipment used to remove organic coatings for structure surfaces must be a direct pressure feed abrasive blasting unit capable of propelling a controlled and continuous stream of plastic media. This equipment requires indicator and regulation devices.

Note: Siphon fed abrasive blasting equipment should not be used for plastic media blasting.

Operator Equipment: Operator safety equipment is required and should include respiratory equipment, blast hood, hearing protection and intercommunication devices. Recommended operator equipment consists of the following:

1. *Breathing air:* The system shall include, as a minimum, all breathing air purification equipment necessary to provide Class D breathing air as described in the Compressed Gas Association Commodity Specification G-7.1-1966. If a compressor is used to supply the air, it must either be a breathing air compressor or a compressor equipped with both a carbon monoxide and high temperature alarm. A breathing air filter shall be provided to remove particulate, moisture, and oil vapor. Clemco Industries Model CPF-80 or equivalent. A carbon monoxide monitor shall be provided. Dynamation, Inc. Model ABL-50 or equivalent.
2. *Voice communication:* Radio headset.

3. *Hearing protection:* The PMB operations must have protection from noise levels exceeding 85 decibels measured on an "A" weighting scale.

4. *Blast suit:* "Apollo" helmet with leather front and suit and gloves. Clemco Industries or equivalent.

Equipment Maintenance: The PMB equipment must be operated and maintained in accordance with the equipment manufacturer instruction manual. The blasting nozzle and plastic media transport hoses shall be visually inspected for excessive wear and replaced as required prior to each use. The nozzle diameter must be checked periodically to verify an exit diameter of 0.50 inch plus or minus 0.05.

Facility Requirements: The PMB Facility must meet the requirements of OSHA 29CFR1910.9. A minimum cross-draft ventilation rate of 75 cubic feet per minute shall be provided. The dust collection system shall be capable of removing 99.97% of the particles 0.3 (three tenths) micron or greater in size from re-circulation systems. The dust monitoring equipment must be capable of continuously monitoring explosive conditions of the air in the facility. **WARNING: Plastic media dust accumulations in a confined area may produce explosive conditions. The air flow should be continuously monitored to ensure that the airborne concentration does not exceed 15% of the minimum explosive dust level inside the PMB room.**

A minimum of 55 foot candles (55 lumens/sq ft) of illumination at the working surface is required. Portable lighting should be provided as required to prevent shadows on the sides and the bottom of the structure. All lighting must meet the requirements of the National Electric Code for Class II, Division I locations, i.e. dust ignition proof.

Compressed air used for the PMB equipment must have a maximum relative humidity of 60% at 70 degrees F (21 degrees C). The air must be filtered to remove moisture, oil and solid particles. Moisture or oil in the compressed air will cause the plastic media to clump, clogging the metering valve and producing erratic performance.

Safety Requirements: The following safety requirements

should be instituted for the PMB Facility: A flashing safety light must operate during PMB operations and must be located outside all PMB room doors. The doors must be designed to open outward. All electrical equipment, i.e., motors, lighting, and outlets must meet all the requirements of the National Electrical Code Class II for the areas in which they are located. Utilities associated with the facility must meet the requirements identified in Article 512 of the 1990 National Electrical Code. Static straps/grounding cords and grounding points are required in the PMB facility as required by the individual structures manual. Integrated emergency lighting is required in the facility to illuminate exits in case of power failure and illuminated exit signs are required inside the facility. A closed-head, wet/dry automatic sprinkler system that complies with OSHA 1910.159 and NFPA 13, Vol I is required in each room and enclosure. Operator must periodically sweep the PMB room walls to prevent dust accumulation. A digital dust concentration monitoring system is required for the PMB room. The air flow should be continuously monitored to ensure that the air borne concentration does not exceed 15% of the minimum explosive level inside the PMB room. The PMB operation must have protection from noise levels exceeding 85 decibels measured on the "A" weighting scale. The noise level outside the PMB room must not exceed the 85 decibel level at a radius of 10 feet from the PMB equipment and enclosure. The reading should be taken at a height of five feet above the floor (ear level).

Preparation of Structure for PMB: Clean the excess dirt and grease or oil from the structure using an appropriate cleaning solution. Mask all areas that are not to be stripped. Seal all openings in the structure in the area to be blasted to prevent intrusion of the plastic media into the interior areas. Cover all transparent plastic and glass surfaces.

Removal of Paint: Position the cleaned and masked structure at the PMB site. Ground the component if necessary. Remove the paint coating from the structure using Plastic Media Type V, Grade A using the following parameters.

For Metallic Surfaces: Air Pressure of 30 +/- 2 psi, Media

Flow Rate of 450 - 480 lbs/hour based on a nozzle diameter of 0.50 inches, Impingement Angle of 15 - 30 degrees, Stand-off Distance of 18 - 24 inches, and Dwell Time of less than 1 second.

For Composite Surfaces: Air Pressure of 18 - 20 psi, Media Flow of 450 - 480 lbs/hour, Impingement Angle not to exceed 15 degrees, Stand-off Distance of 24 - 30 inches, and Dwell Time of 0 seconds.

The operator should keep the nozzle moving at all times and maintain the stand-off distance, pressure and impingement angle (angle of attack) specified. This will enable the removal of one layer of coating at a time from the surface and prevent and damage to the substrate material. Remove all plastic media dust and residual dust by use of a vacuum cleaner.

REFERENCES

4-1 Westerman, E. A., and Roll, P. E., "An Apparatus To Prepare Composites for Repair," 34th International SAMPE Symposium, pp. 1041-1051 May 8-11 (1989)

4-2 Wegman, Raymond F., "Surface Preparation Techniques for Adhesive Bonding," Noyes Publications, Park Ridge, NJ (1989)

4-3 MIL-HDBK-337, Military Standardization Handbook: Adhesive Bonded Aerospace Structure Repair

5

TOOLS, EQUIPMENT AND FACILITIES

COMPOSITE REPAIRS

The cutting of advanced composite materials - fabrics, prepregs, and laminates - requires special tools. The inherent toughness of these materials is derived from the combination of their physical properties and their organic nature, To obtain optimum results, a shearing action using sharp tools with low heat generation is desirable. The cutting of the outermost plies requires particular attention. The fibers of the outermost ply must be restrained by firm external backing or the cutting action must be directed toward the interior of the composite. Rotary tools such as drills and routers must be designed so that the fabric is cut into very short strands which will not wrap around the tool. The cutting surface of the tools must be kept clean. Adherence of resin to the cutting surfaces often causes apparent dulling of the cutting edge. Table 2-1 contains a list of some special tools for cutting composite materials. Some suppliers of the tools are given in Table 2-4. Suggested operating conditions for machining composites containing Kevlar are given in Table 2-3.

TOOLS FOR CUTTING COMPOSITE FABRICS AND LAMINATES

Tools for Cutting Prepreg and Fabric Materials

Hand Shears Serrated - these are specially ground.

Hand Shears Conventional - have limited wear life and must be sharpened frequently.

Power Shears - special cutting edges.

Rotary Cutter - special blades.

Razors/Knives - standard or tungsten carbide blades.

Ultrasonic Cutter - ultrasonically vibrated.

Die Cutters - serrated band steel dies.

Tools for Working on Composite Laminates

Drill Bit - Modified Twist Drill (sickle shape-brad point): This bit's design has 2 sharp positive peripheral cutting edges in sickle form, so as to load the reinforcing fibers in tension while cutting them in shear. The center point is not too prominent to avoid delamination. The drill has stable cutting edges to minimize breakage. Tungsten Carbide is preferred for long tool life, the P-Carbide types for hand tools and the K-Carbide types for drill-presses and numerically controlled machines. The operating speed is dependent upon the composite type; for thin laminates (less than 0.060") the speed should be between 12,000 and 20,000 or higher RPM, for medium laminates (0.060-0.160") the speed should be between 10,000 and 16,000 RPM, for thick laminates (greater than 0.160") the speed should be between 8,000 and 12,000 RPM. For sandwich structures use a speed between 12,000 and 20,000 RPM. Feed at 0.002 inches per rev. See Figure 5-1.

Drill Bit-Single Flute Drill: Recommended only when modified twist drill can not be used: e.g., for Kevlar/metal stack-ups or other extremely hard materials. See Figure 5-2.

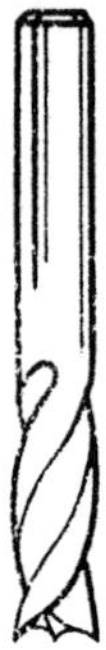

Figure 5-1: Modified twist drill. (sickle shape-brad point).

Figure 5-2: Single flute drill. (Boeing Patent)

Drill Bit - Spade drill without serrations: Use with drill press or with guide only. Tip has a 15-20 degree angle. Tungsten carbide preferred. Recommended for thin laminates when the sickle-shaped brad point modified twist drill can not be used. For very small holes only, less than 0.125" diameter or for economical reasons (a cheap alternative when only a few holes need to be drilled).

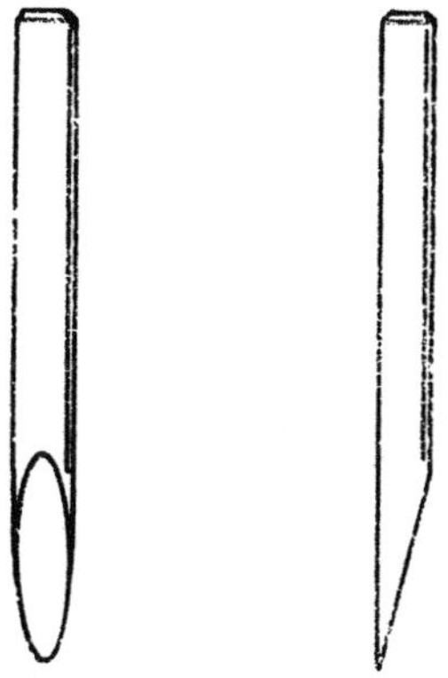

Figure 5-3: Spade drill.

Drill Bit - Self-centering drill: A single point, self-centering serrated drill which combines up and down cutting flutes. May be used where it is difficult or impossible to access the exit side of the material to provide backup support, or where other considerations prevent use of tools already discussed. This highly specialized tool requires considerable skill to produce a clean, properly dimensioned hole and is not recommended for general use. To obtain clean entrance and exit surfaces, this drill must rapidly penetrate the work piece until the centerline of the bit is approximately mid-way between the entrance and exit surfaces of the laminate. After initial penetration, and while continuing to operate the drill motor, the tool is moved rapidly in an orbital manner about the center line for several seconds allowing the cutting edges to shear the uncut fibers toward the center of the laminate thickness. During this process radial movement of the tool must be minimized to insure a properly dimensioned hole. This tool is not recommended for use with low resin content composites or for drilling into underlying metal support.

Router Bit - Split helix: Tungsten Carbide with 2-4 helixes (straight grooves) for up and down cutting. 1/4" for pin router tables and hand routers, 1/2" for numerically controlled routing machines.

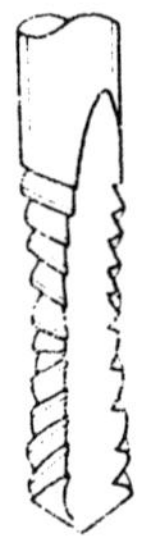

Figure 5-4: Self-centering single point drill.

Figure 5-5: Split helix router bit.

Router Bit - Spiral shear router: Router consists of a cutting spiral which operates against a shear busing. The tool is installed in a special nose with a guiding device which fits in to a pneumatic hand motor with variable speeds. This router is recommended for use with thin laminates (up to 0.060").

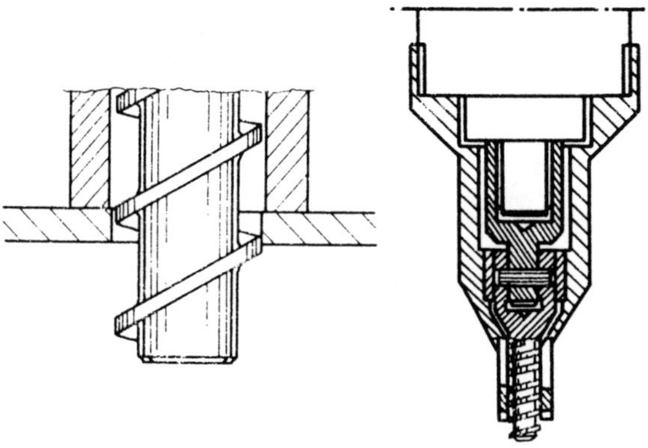

Figure 5-6: Spiral shear router.

Router Bit - Opposed helical routers: This router bit is not recommended for cutting laminates containing Kevlar. It will require some finishing of the secondary surface.

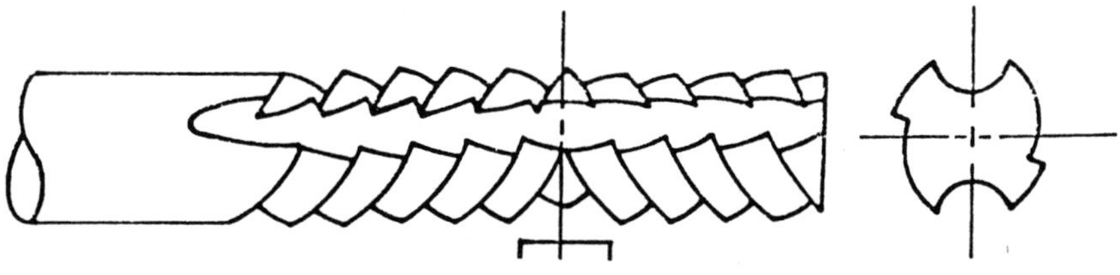

Figure 5-7: Opposed helical router.

Countersinking Tool: This tool should have 1 or 2 sickle shaped cutting edges with a positive rake in the radial direction to avoid fuzz at the periphery, and a slightly positive rake in the axial direction for aggressive cutting action and smoother surface. If less aggressive cutting is desired, a slightly negative rake in the axial direction if used. The cutting edge is designed to generate proper chips (no powder) and to remove them quickly. A micro stop is required for hand-machining to control hole depth.

Figure 5-8: Countersinking tool.

Counterboring Tool: This tool has 2 very sharp curved peripheral cutting edges (vertical cut) and 2 very sharp, straight radial cutting edges with slightly negative rake (horizontal cut). There is 0.002" difference between peripheral and radial cutting edge and pilot. The tool must fit tightly in the pre-drilled hole. Tool must be used with an microstop for hand-machine.

Saber Saw Blade: High speed shank with alternating sets of opposing teeth designed to produce a fuzz-free cut on Kevlar laminates up to 1/2" thick.

Band Saw Blade: Fine tooth blade with 14 to 22 teeth per inch (mounted in the saw in reverse for Kevlar laminates so that the heel of the tooth enters the composite first). Either a raker or straight set may be used. Teeth should be honed. Honing may be done while the blade is turning, using a standard shop mechanic's honing stone 3/4 inch by 2 inches by 6 inches. Apply hone to side of the band to form the teeth shown in Figure 5-11. High speed should be used (4500 to 6500 SFPM) with cooling if possible.

Figure 5-9: Counterboring tool.

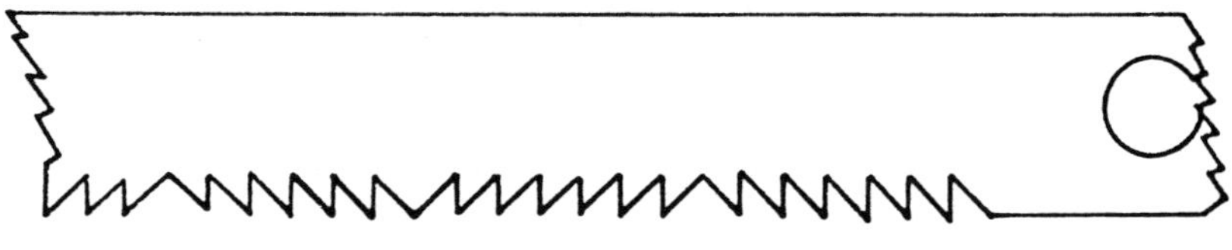

Figure 5-10: Saber saw blade.

Circular Saw Blade: A metal slitting saw 3 inches in diameter, 1/8 inch wide with 28 staggered teeth can be used to cut Kevlar laminates. Diamond wheels are used to cut glass laminates.

Hole Saw: Tungsten carbide grit-edged hole saws may be used to cut laminates. Fuzzing at the entrance and exit surfaces may be reduced by using external surfaces of aluminum, brass, "micarta", or fiberglass laminate. The inside wall of the hole will be smooth.

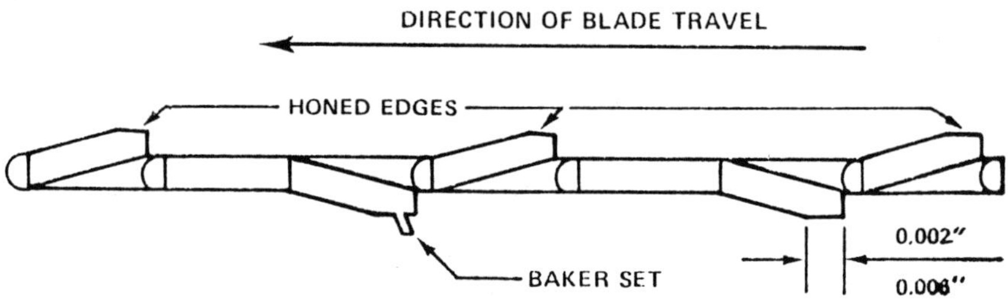

Figure 5-11: Band saw blade.

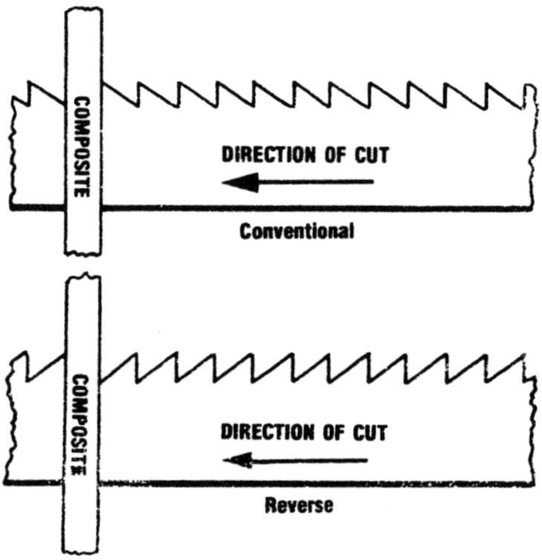

Figure 5-12: Reverse band sawing of Kevlar laminates.

Fly Cutter: Large holes (1.0 to 12.0 inches) may be cut with a fly cutter with special rakes and angles depending upon the laminate material. For Kevlar, the cutting tool must be fixed in such a way that the cutting edge lies backwards so as to pull the fiber toward the middle part of the hole.

Mechanical Scarfing Apparatus (MSA): The MSA permits the rapid, high-quality preparation of a scarf joint in damaged laminates. The quality of the scarf will be several orders of magnitude better than hand scarfs. The apparatus allows precise control of bondline thickness. Scarfing time can be cut by 50% or more as compared to hand scarfing. The tool can be operated by semi- skilled personnel with minimum instructions. The MSA is easily portable and can be transported, set up and operated by one person. The Mechanical Scarfing Apparatus is shown in Figure 5-13.

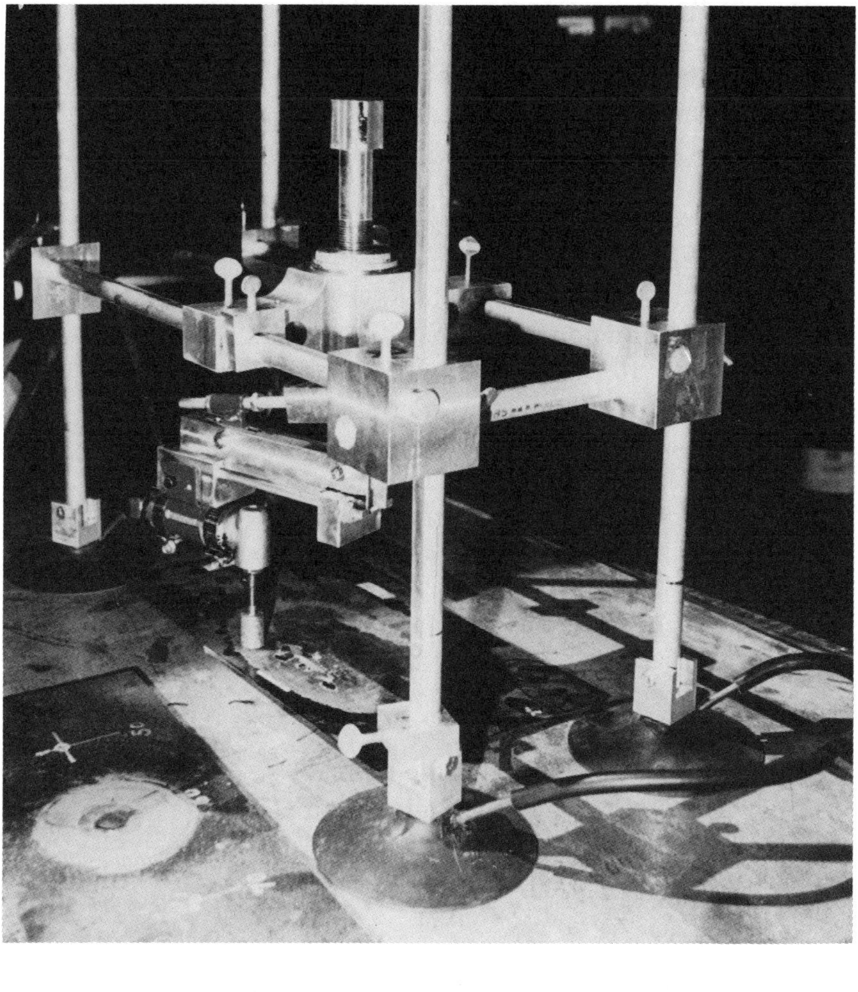

Figure 5-13: Mechanical scarfing apparatus (**Boeing Patent**).

Tool Kits: Special tool kits are available which are designed to provide a complete solection of the tools needed to perform on-location repairs. Kits of this type have been designed for use on metal, Kevlar, graphite and fiberglass materials. Replacement parts are available. An example of this type of kit is the ATACS Tool Kit No 8601, this kit contains the following components.

1 ea. Pneumatic Hydro Check Feed Control Assembly
1 ea. Pneumatic Drill 2,300 RPM
1 ea. Pneumatic 18,000 RPM 90 Degree Grinder
1 ea. Pneumatic 18,000 RPM Router Assembly
1 ea. Pneumatic Bench-Vac
2 ea. #53 Carbide Tipped Drill Bit 0.0595" Dia.
3 ea. 0.2495 X 3" Braze Diamond Router 60 Mesh
3 ea. 0.2495 X 2-1/4" Braze Diamond Router 60 Mesh
1 ea. 1" Diamond Hole Saw 1/4" Pilot
1 ea. 1-1/2" Diamond Hole Saw 1/4" Pilot
1 ea. 2" Diamond Hole Saw 1/4" Pilot
1 ea. 1/8" 100 Degree Countersink
1 ea. 5/32" 100 Degree Countersink
1 ea. 3/16" 100 Degree Countersink
1 ea. 1/4" 100 Degree Countersink
3 ea. 1" X 1" X 2-1/4" Brazed Diamond Router 60 Mesh
1 ea. 2" Putty Knife Style Core Cutter
1 ea. Aluminum Cutter 1/8" Bit
1 ea. Core Knife
1 ea. 2" ROLOC Disc Holder
1 ea. 1" ROLOC Disc Holder
1 ea. Aluminum Cutter 1/4" Bit
1 ea. Barrel Grider Attachment
1 ea. 1" Barrel Shape Grinder Stone
1 ea. 1" Mushroom Shape Grinder Stone
1 ea. Countersink Feed Cage/Holder
1 ea. 1" Honeycomb Shaver
1 ea. 2" Honeycomb Shaver
1 ea. Style K Double Helix Laminate Cutter
1 ea. 1/8" Spade Drill

1 ea. 3/32" Spade Drill
1 ea. 3/16" Spade Drill
1 ea. 1/4" Spade Drill
1 ea. 0.2510" Reamer
1 ea. 0.1885" Reamer
1 ea. 0.1572" Reamer
1 ea. 0.1260" Reamer
2 ea. 0.2340" 1/4 X 6" Webb Thinned Flute
4 ea. 1/8" Cleco with 1/4" Wing Nut
4 ea. 5/32" Cleco with 1/4" Wing Nut
4 ea. 3/16" Cleco with 1/4" Wing Nut
4 ea. 1/4" Cleco with 1/4" Wing Nut
12 ea. 1" 80 Grit ROLOC Discs
12 ea. 2" 80 Grit ROLOC Discs
12 ea. 1" 180 Grit ROLOC Discs
12 ea. 2" 180 Grit ROLOC Discs
12 ea. 20 Grit Barrel Type Grinder Core
12 ea. 180 Grit Barrel Type Grinder Core

The ATACS Tool Kit is shown in Figure 5-14.

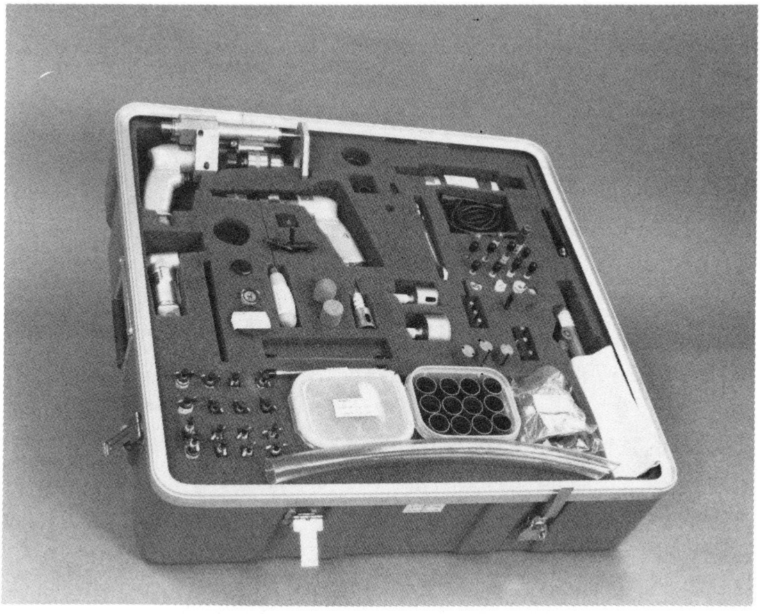

Figure 5-14: Atacs Tool Kit No.8601

EQUIPMENT

Hot Bonding Equipment

Controllers: Hot bonding control units contain the temperature control system, monitoring systems and the vacuum source. The units are available in single and multiple zones. They can be programmed to control the temperature rise time, dwell time and cool down time. Multiple step cures can also be programmed into the controller. Typical control units are shown in Figures 5-15 and 5-16.

Heat Blankets: Heat blankets are silicone rubber pads which have embedded electrical They come in custom sizes and shapes to fit the repair. The blanket used should be 2 inches larger than the repair area. The electrical requirements are 115 volts, 60 cycle, 5 to 10 watts per square inch, the amperage should not exceed 30 amps per zone.

Vacuum Bag Valves and Base Plates: Vacuum valves and base plates come in a variety of shapes and sizes. Examples of these are shown in Figure 5-17.

Heat and Vacuum Tables: The heat and vacuum table is designed to offer maximum efficiency for part repair. The table can be obtained with or without a top laminating hinged lid. The table can be controlled by use of a contoller of the type described earlier or by a self contained control unit and pump. A heat and vacuum table is shown in Figure 5-18.

PROCESSING FACILITIES

Processing facilities should meet those requirements specified in MIL-A-83377.

Controlled Atmospheric Layup Areas

All activity in the controlled atmospheric layup area is done by individuals qualified or certified in the techniques of adhesive bonding. All work requires white gloves and no contamination.

Figure 5-15: Single zone controller.

Structural film adhesives and prepreg materials are stored in freezers in accordance with the manufacturer's instructions until 24 hours before use. When the material is removed from the freezer, it must be allowed to equilibrate to room temperature in the controlled atmospheric area before opening to prevent moisture condensation on the material. The material can be conveniently placed on a roll stand so it can be unrolled onto a cutting table and cut to the size and shape required.

General Criteria For The Layup Area: The requirements for the layup areas of adhesive bonding shops engaged in repair of structure should be as follows:

1. Limits of airborne particulate matter, temperature, and humidity should be as specified in MIL-A-83377.
2. Airborne hydrocarbons will be controlled by imposing operation restrictions. The number used for identification of each class corresponds to the maximum number of particles, 0.5 micrometers or larger, permitted in one cubic foot of air.

Figure 5-16: Multiple zone heat controller.

3. Controlled environment areas for bonding do not fall within the definition of "clean rooms" according to FED-STD-209. The requirements for a "controlled area" room are recommended as minimums. However, FED-STD-209 should be used as a guide for facilities construction.

4. The adhesive and prepreg application area should be a separate, enclosed area. Special facilities ensure that the application area is dust controlled. Filtered air constantly forced into the area maintains air pressure at a higher level than the surrounding area. This prevents contaminating substances from entering the dust controlled area when the doors to the surrounding area are opened. Unless otherwise noted, temperature and humidity are monitored and meet the following requirements: Temperature: 60 to 80 degrees F. (15.5 to 27 degrees C.). Humidity: 70% RH maximum. Any method and equipment may be used for monitoring that gives assurance of maintaining proper control.

5. Tables or work benches used for cutting, layup, core splicing, etc., should be covered with a minimum of 1/2 inch thick polyurethane sheeting. All other work tables should be covered with a suitable material to prevent damage to any component being bonded.

6. No steam heated or gas operated vehicles or equipment should be allowed in the controlled area.

7. Smoking or eating will not be permitted in the controlled area.

8. Unpacking of materials, filling, drilling, cutting, grinding, or machining of metals or composite materials must not be permitted in the controlled area.

9. All materials, parts, tools, equipment, and other items to be used in this area must be free of extraneous dirt, grease, oil, talc, wax, etc. before it is brought into the controlled area.

10. Electrically operated equipment may be allowed in the controlled area for short periods of time only.

Figure 5-17: Types of vacuum valves and base plates.

Operating Conditions

1. Restrict all unnecessary traffic.
2. The list of personnel approved for access shall be screened.
3. Cleanliness of parts entering the area shall be checked.
4. Protective clothing requirements shall be met.

NOTE: Protective clothing must be worn for the purpose of limiting airborne particulate matter generated by the personel. The use of protective clothing also promotes a "clean work psychology".

Area Maintenance

1. Areas should be constructed to be easily maintained.
2. Areas should be partitioned to help prevent drafts from carrying contaminants in from the noncontrolled areas.

Personnel

1. Personnel, including supervision and maintenance, that work in the controlled environments need appropriate training in clean room or controlled area procedures.
2. The personnel working in these areas should be certified in clean room or controlled area procedures.
3. Authorized observers should be familiarized with pertinent procedures and, when entering the area, must be accompanied by a certified person. Their presence should be restricted to the aisle zones.

Inspection

1. Unless otherwise designated, the quality control organization is responsible for ensuring that the controlled area meets all criteria required for high quality bonding operations.
2. The amount or level of airborne particulate matter is determined by either the fallout, volumetric, or automatic particle counter method at the option of quality control.

CURING EQUIPMENT

Autoclaves

The purpose of the autoclave is to provide a means of applying uniform high pressure and temperature to a part for curing. It has definite advantages over the use of other methods such as vacuum bags and heating blankets in that the higher pressure provide superior quality bonds. The detail parts are better consolidated for intimate bondline contact.

Autoclaves come in various sizes. Heat is provided by electricity, steam or hot oil, with internal air circulation. Pressurization is achieved by positive air pressure or inert gas. The assemblies to be pressurized are sealed under a bagging

Figure 5-18: An Atacs Model 8742 Heat and Vacuum Table.

film and either vented to the outside atmosphere or connected to a vacuum line. It is important that pressure not be directly applied to honeycomb assemblies without the film barrier, as the pressure will laterally crush the core. Conventional practice is to cover and seal the repair components after assembly for bonding. A vacuum is then pulled on the assembly to hold the details in position until the part is moved into the autoclave. Approximately 10 psi positive pressure is then applied to the autoclave before the vacuum lines are vented to the atmosphere. Vacuum systems should be constructed so that they can be vented to the atmosphere and connected to a guage system on the control console. Venting the vacuum outlet lines prevents a differential pressure from accumulating under the bag film. By monitoring the vacuum guages the operator has control of the go/no-go cure cycle for any assembly.

Differential pressure from 0 to 6 inches Hg (0 to 3 psi) is allowed in the sealed area during a cure cycle. If the differential pressure rises above the 6 inch Hg (3psi) maximum, provided the temperature is 150 degrees F or less, the cycle can be aborted and the leaks or other malfunction repaired before continuing the cure cycle. NOTE: The above applies to the use of adhesives which cure at 250 degrees F or higher.

A typical conventional autoclave vacuum system is shown in Figure 5-19. Care must be exercised in the placement of part in the autoclave so as not to cause a phenomenon called shadowing. Shadowing occurs when part are improperly placed in the autoclave and can cause cool spots as can be seen in Figures 5-20 and 5-21.

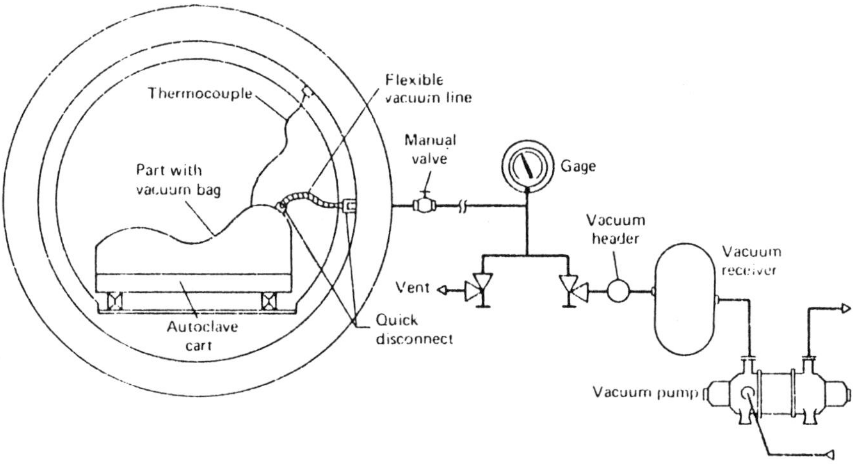

Figure 5-19: Conventional autoclave vacuum system. (Ref. 5-1)

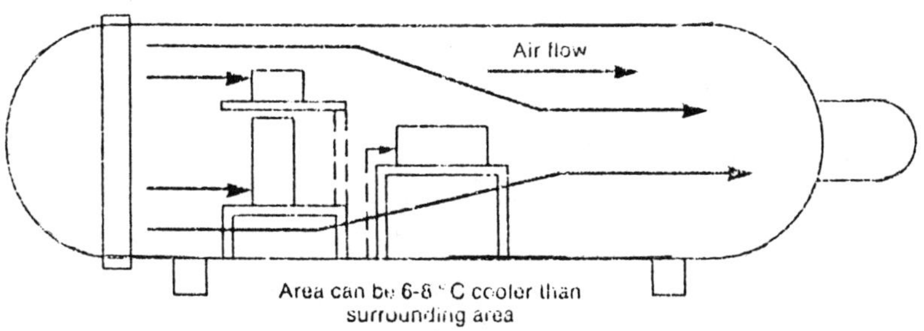

Figure 5-20: The effect of shadowing caused by improper autoclave loading. (Ref. 5-2)

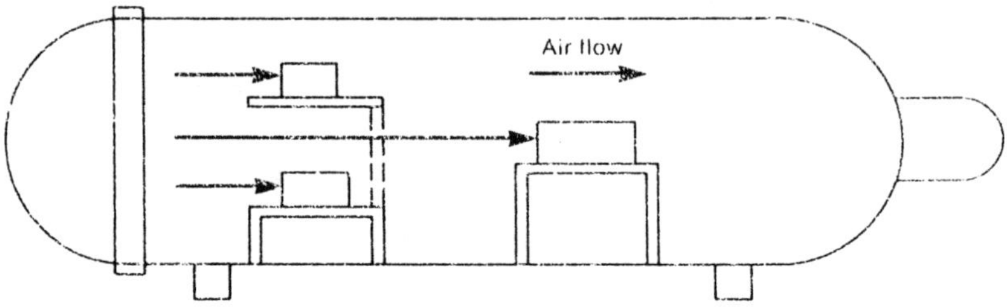

Figure 5-21: Correct autoclave loading to prevent shadowing. (Ref. 5-2)

Autoclaves should be operated by trained and certified operators whose task it is to see that all safety procedures are followed throughout each cycle. Automatic recording and controlling equipment is available and should be used with each autoclave. The autoclave should be equipped with multiple thermocouple recording and/or controlling equipment so actual part temperatures can be monitored during the cure.

The curing of composite materials in an autoclave should be done using inert gases in stead of air for pressure, since these materials are combustible. The inert gas (carbon dioxide or nitrogen) may be produced by an exothermic gas generator. The exhaust gases are typically transferred to a compressor system where the gases are compressed and stored in pressure vessels of adequate capability for the particular autoclave.

Curing Ovens

Curing ovens range in size from the very small to the walk-in types. Larger ovens are typically gas heated with an air circulation system. Ovens should be equipped with a vacuum supply and multiple thermocouple connectors. These should be coupled with automatic controlling and recording equipment. Ovens may be used for initial cures when vacuum pressure is used. They may also be used for post-curing when the initial cure is accomplished in an autoclave under higher pressure.

Vacuum Bagging of Structures Under Repair

Bagging is the standard terminology referring to the sealing of a repair area by encapsulating it in an impervious plastic film during the cure cycle. Consolidation pressure may then be effectively applied to the repair either by pulling a vacuum on the sealed area or by applying pressure to the part in an autoclave and venting the sealed area to the external atmosphere. For the vacuum cure, heat may be obtained using heat blankets and a controller or by placing the assembly in an oven. In an autoclave, heating is an integral part of the system. Where the equipment is available, the autoclave cure is generally recommended.

The following discussion on vacuum bagging is directed toward the use of the vacuum curing system. Satisfactory bonds can be made using the vacuum system, however, certain precautions need to be taken. The detail parts should fit well. There may be a tendency for the vacuum to laterally move the honeycomb cell walls if care is not exercised. In general, the following guidelines apply:

1. Limit the vacuum on partially bagged honeycomb components or on assemblies having light core and thin skins to 8 to 10 psi.
2. A full vacuum may be used for envelope (i.e., fully) bagged assemblies that have thick skins and heavy density core.
3. When the assembly is to be left under vacuum for an extended period of time, (e.g., overnight), reduce the pressure to 5 to 6 psi.
4. When curing a prepreg material or wet layup, use a caul plate which is larger than the repaired area to evenly distribute the pressure.

Several approaches may be taken to bag the part. If the area being repaired is small and the part is quite large, it may be easiest to seal only the area around the repair. In other instances it may be necessary to envelope bag the entire part. For most repairs in this handbook, the area bagging technique is used and will be described in a subsequent section.

The following terminology is generic to the bag cure process:

Bagging: The application of a specific group of materials necessary to seal off a repair area from the ambient environment. This procedure is essential to obtain a bonding pressure. It is achieved by drawing a vacuum or pressurizing the assembly in an autoclave.

Bagging Film: An impervious plastic film (e.g., nylon, PVC, Mylar, etc.) that covers the repair area or completely envelopes the entire assembly. It is usually 2 to 4 mils in thickness and is discarded after each cure cycle because the plasticizers in the film are affected by heat, causing cracking and leaks.

Bleeder or Breather Cloth: A porous cloth to allow the passage of air or gases and to absorb excess adhesive flash. Air and gases flow over or around the assembly or repair to any provided outlet such as vacuum probes, or vacuum bag valves.

Caul Plate: A support plate that is used to distribute pressure. It may be used over a thin skin that otherwise might dimple into the core cells. It should be used with wet layup composite structures. It may be used to bridge over an area where pressure is not desired. It may also be used with an assembly tool to fix part details in place.

Release Film: Thin films produced from nonstick materials, such as FEP, teflon, modified halohydrocarbons, PVF, etc. They are interleaved between any adhesive system or sealant and the surface not intended to be bonded.

Vacuum Probes: Also called vacuum bagging valves. The probe provides air passage through the bagging film. It provides a hose connection for the vacuum line. A number of probe designs exist and are commercially available. Normally the probe base is under the bag film and a rubber or silicone disc on the exterior surface acts as a seal.

Vacuum Bag Sealant: Also known as vacuum seal putty. A semi-cured plastic material having tackiness for holding and sealing the bagging film during the cure cycle. This is the sealing medium between the bagging film and the part or tool base.

Vacuum Bagging Sequence

General: In order to apply a vacuum bag to a repair the following materials and equipment should be available:

1. Release film
2. Breather Cloth
3. Bagging film
4. Vacuum bagging sealant
5. Heater blanket
6. Vacuum bag valves and base plates (2)
7. Vacuum guage
8. Vacuum hose with quick disconnect
9. Vacuum pump
10. Temperature controller
11. Thermocouples (4)
12. Masking tape (high temperature)
13. Electric power (120V)
14. Scissors.

Application:

1. Cover the repair with a sheet of perforated release film which extends approximately 1 inch beyond the repair area.
2. Place 4 thermocouple (TC) leads on the center of the repair area. The TC leads may be placed directly on the repair if the skin is metallic or on the metal caul or pressure plate if the repair has a composite skin. A piece of masking tape may be used to hold the TC leads in place.

3. Cover the release film and the TC leads with a sheet of breather cloth. Small pieces of tape may be used to secure the breather cloth.
4. Cover the breather cloth with a sheet of perforated release film.
5. Place the heater blanket over the repair area, use a heater that is 2 inches larger than the repair area.
6. Cover the heater blanket with 2 to 4 layers of breather cloth to help retain the heat at the repair surface.
7. Construct a frame of vacuum bagging sealant around the covered area.
8. Lay the TC leads and the heater blanket leads on top of the vacuum bagging sealant, remove the separator paper from the sealant in this area, take care to separate the leads. Cover the leads with an additional layer of vacuum bagging sealant. Press the two layers firmly together.
9. Place the 2 vacuum valve base plates on the breather cloth at opposite edges of the repair area.
10. Cut small crosses in a sheet of vacuum bagging film over the center or the base plates and insert the male ends (ends containing the roll pins) of the vacuum valve through the crosses, folding the cut ends back. Attach the valves to the base plates and secure by turning the knurled part of the valve.
11. Remove the separator paper from the vacuum bagging sealant in sections and attach the bagging film to the sealant, using pleats or "rabbit ears" to build in flexibility. NOTE: The pleats or "rabbit ears" are made by using a 5 to 6 inch strip of vacuum bagging sealant attached to the underside of the bagging film at spots along the frame of sealant. (Figure 5-22.) The bagging film and the sealant are folded to form 1 to 2 inch pleats, the rest of the strip is then pressed to the frame of sealant.

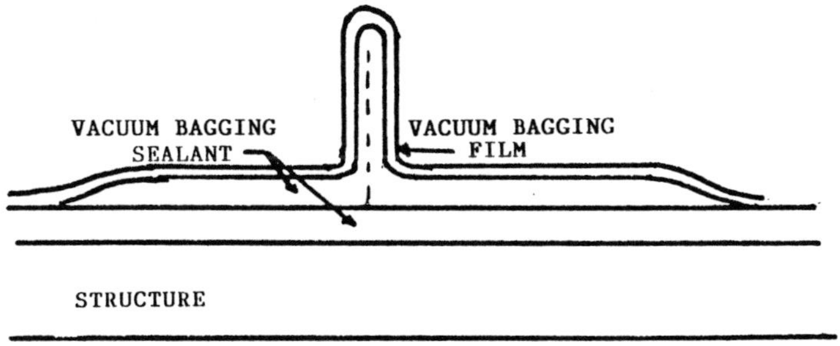

Figure 5-22: A pleat or "rabbit ear" in a vacuum bag.

12. After the bag has been secured on all sides, press the film firmly to the sealant. A smooth hard material such as the handle of the scissors slid along the surface over the sealant helps to secure the film against leaks.
13. Attach the vacuum hose to one vacuum valve and the vacuum gauge to the other valve.
14. Attach the vacuum hose to the vacuum pump and turn on the vacuum. Check for leaks, a good vacuum should hold a minimum of 22 inches of Hg.
15. Plug the heater blanket and the TC leads in to the controller. Plug the controller unit in to the power source and turn the unit on.
16. Set the program for the cure cycle desired and cure in accordance with the applicable instructions.
17. Cool paste type adhesives to below 90 degrees F. and structural film adhesives to below 140 degrees F. before removing the vacuum bag.

REFERENCES

5-1 Engineered Materials Handbook, Volume 1-Composites, ASM, page 646, (1990).

5.2 Ibid, page 747.

6

FIELD LEVEL REPAIR OF COMPOSITE STRUCTURES

This chapter provides instructions for making repairs on adhesive bonded composite faced and composite laminate structures. These repairs are essentially patch type repairs and limiting the use of these methods is the responsibility of the cognizant engineering agency for the structure being repaired.

COMPOSITE LAMINATE STRUCTURES (NON STRUCTURAL)

Method No. 6.1: Repair of Non-structural Composite Laminates Containing Minor Damage

Structures to be repaired by this method must not be primary load bearing structures. The clean out area should be at least two inches from the panel edge.

*Method of Repair. (See **Figure 6-1**):*

1. Mask off an area approximately four inches larger on all sides that the area to be repaired.

2. Remove the paint from both sides using an appropriate method (See Chapter 4.)

3. Remove the damaged laminate using a suitable tool (See Table 2-1.)

4. Mix the laminating adhesive (See Table 2-1) in accordance with the manufacturer's instructions.

5. Apply two piles of glass cloth impregnated with the laminating adhesive to the external side of the laminate. The first ply should overlap the damaged area by one inch. The second ply should overlap the first ply by one inch, thus achieving a one inch step on all sides.

6. Cover the repair with a sheet of release film and a 1/8 inch minimum thickness aluminum plate in a vacuum bag and cure in accordance with the adhesive manufacturer's instructions.

7. Mix the repair adhesive (See Table 2-1) in accordance with the manufacturer's instructions.

8. Fill the cavity over-flush with repair adhesive and cover with masking tape.

9. Cure the adhesive in accordance with the manufacturer's instruction.

10. Sand both sides smooth feathering the edges of the glass plies and refinish as necessary.

Figure 6-1: Repair of non-structural composite laminates containing minor damage.

Method No. 6.2: Repair of Non-structural Composite Laminates Containing Major Damage

Structures to be repaired by this method must not be primary load bearing structures. The clean out area should be at least two inches from the panel edge.

*Method of Repair. (See **Figure** 6-2):*

1. Mask off an area approximately four inches larger on all sides than the area to be repaired.
2. Clean the masked off area with a trichloroethane dampened clean white cotton cloth and remove the paint using an appropriate method (See Chapter 4.)
3. Layout the area to be removed. The layout area should have 1/8 inch minimum radius corners.
4. Drill four 1/4 inch minimum diameter holes in the corners. (See Table 2-1 for special drill bits).
5. Cut the laminate between the drilled holes. (See Table 2-1 for special saw blades). (See Figure 6-2, View A.)
6. Prepare a metal plate of 0.40 inch minimum thickness to act as a back-up tool. The plate should be at least one inch larger on all sides than the repair cut-out. Apply a non-silicone release agent to the surface of the tool.
7. Position the back-up tool over the hole on the back side of the laminate. Seal the edges with vacuum bagging sealant. (See Figure 6-2, View B.)
8. Cut an insert of the same laminate material which is approximately 0.03 inch smaller on all sides than the hole in the original panel.
9. Mix the two part paste adhesive (See Table 2-1) in accordance with the manufacturer's instructions.
10. Apply the adhesive to all sides of the insert and the cavity. Place the insert into the cavity. Fill any voids with the paste adhesive. (See Figure 6-2, View C.)

11. Allow the adhesive to cure in accordance with the manufacturer's instructions.

12. If laminate material is not available to fabricate the insert, cut cloth plies of the same material, size and number of plies as the original laminate. Impregnate the plies with the laminating adhesive and fill the hole with the impregnated plies. Cover the repair with a sheet of release film. Apply a vacuum bag and cure under vacuum in accordance with the manufacturer's instruction.

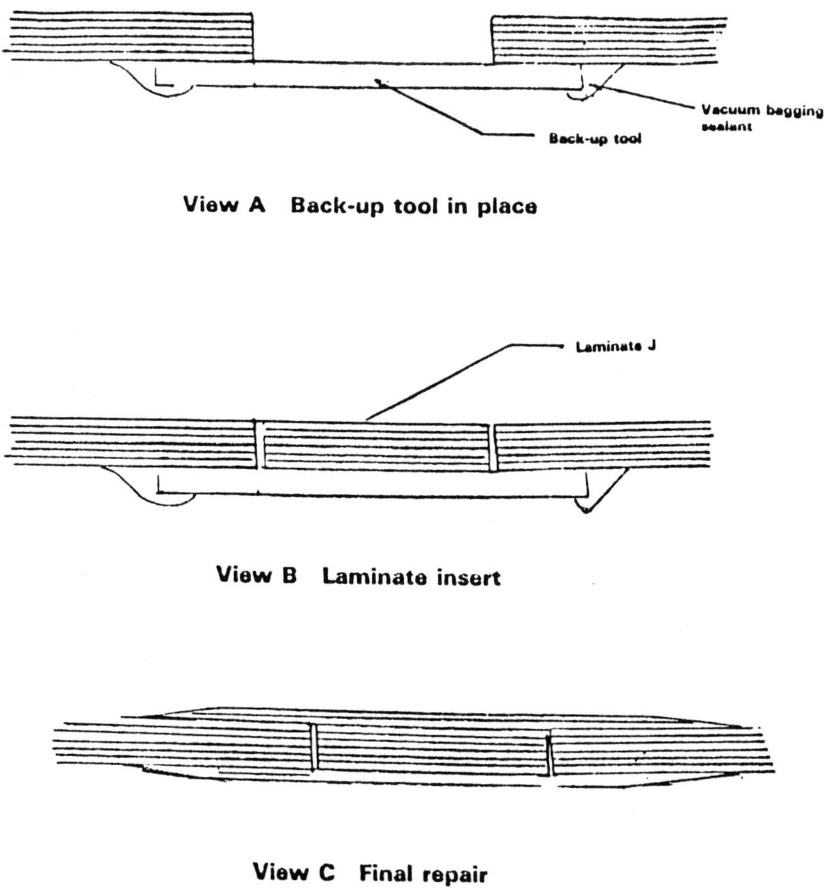

Figure 6-2: Repair of non-structural composite laminate containing major damage.

13. Sand off any excess adhesives.
14. Mix the laminating adhesive and apply two plies of impregnated glass cloth over the repair. The first ply should overlap the repair by one inch on all sides, the second ply should overlap the first by one inch on all sides.
15. Cover the repair with a sheet of release film. Place an aluminum plate (1/8 inch thick, minimum) which is approximately 1/4 inch larger than the last ply over the repair.
16. Apply a vacuum bag and cure under vacuum in accordance with the manufacturer's instructions. NOTE: Steps 14 and 15 may be accomplished on both sides at the same time or at two separate times.
17. Sand the surfaces smooth and feather the edges. Refinish as required.

REPAIR OF STRUCTURAL LAMINATES

Method No. 6.3: Repair of Structural Laminates Using The Precured Patch Concept

This repair may be used for primary load bearing composite laminate structures. The limits for use of this repair must be determined by a structures engineer for the structure to which it is to be applied. NOTE: The laminate should be thoroughly dried prior to the application of the adhesive to prevent blistering during heat cure.

Method of Repair:

1. Mask off an area which is six inches larger on all sides than the area to be repaired.
2. Clean the masked off area with a trichloroethane dampened clean, white cotton cloth, remove paint using a method from Chapter 4.

3. Remove the damaged material using a suitable tool. (See Table 2-1 for special tools.)

4. Prepare an insert of the same material as the structure being repaired. The insert should be slightly smaller than the hole to allow for a layer of adhesive film to be placed on the edges of the insert.

5. Prepare five pieces of precured laminate. Each piece should be smaller than the preceding piece. (See Figure 6-3.)

6. Prepare five pieces of structural film adhesive (See Table 2-1) the same size as the laminates prepared in Step 5.

7. Assemble the adhesive and laminate over the insert starting with the largest and ending with the smallest.

8. Place the repair in a vacuum bag and cure in accordance with the manufacturer's instructions.

9. Remove the vacuum bag and sand the patch to feather the edges, refinish as necessary.

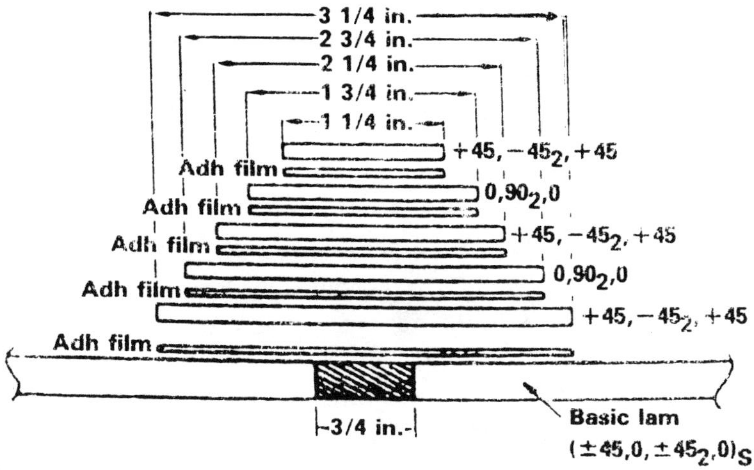

Figure 6-3: Repair of a structural composite laminate using the precured patch concept. (Ref. 6-1)

Method No. 6.4: Repair of Structural Laminates Using The Cure-in-Place Concept

This repair may be used for primary load bearing composite laminate structure. The limits for use of this repair must be determined by a structure engineer for the structure to which it is to be applied. NOTE: The laminate should be thoroughly dried prior to the application of the prepreg to prevent blistering during heat cure.

*Method of Repair. (See **Figure** 6-4):*

1. Mask off an area approximately six inches larger on all sides than the area to be repaired.
2. Clean the masked off area with a trichloroethane dampened clean, white cotton cloth, remove paint using a suitable method from Chapter 4.
3. Remove the damaged material using a suitable tool. (See Table 2-1.)
4. Prepare an insert of the same material as the structure to be repaired. The insert should be slightly smaller than the hole to allow for a layer of film adhesive to be placed on the edges of the insert.
5. Place the insert into the hole.
6. Cut a piece of structural film adhesive that is 6.5 times the size of the cutout.
7. Prepare four (4) pieces of prepreg which increase in size by 1/4 inch larger than the cutout. Prepare an additional four (4) pieces which increase in size by 3/8 inch and two (2) pieces which increase by 1/2 inch. The orientation of the plies should vary in pattern similar to the original laminate. (See Figure 6-4)
8. Remove the separator sheet from one side of the adhesive and center it over the repair insert. Remove the other separator sheet from the adhesive.

9. Place the pieces of prepared prepreg over the adhesive in order of increasing size.

10. Place the repair into a vacuum bag and cure in accordance with the prepreg manufacturer's instructions.

11. Remove the vacuum bag and refinish the repair as necessary.

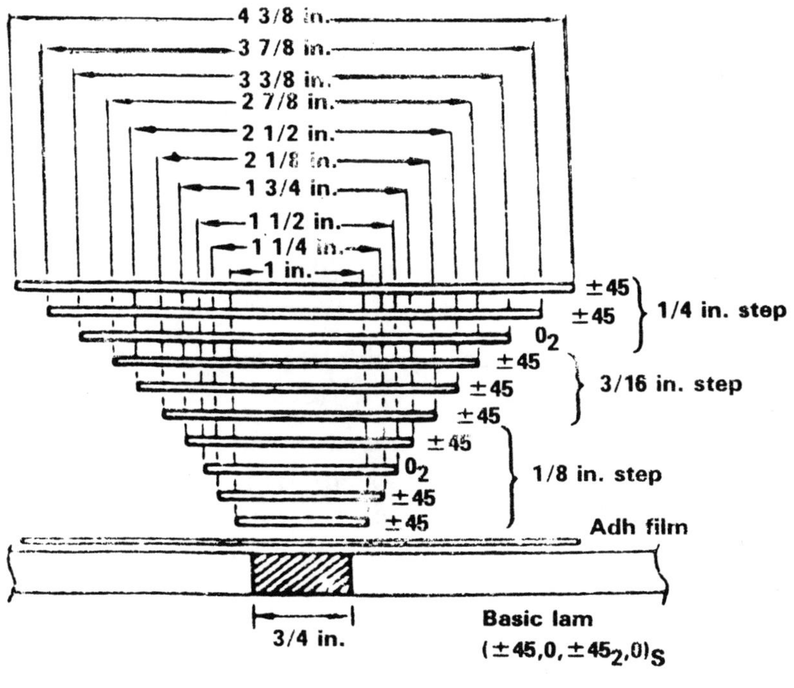

Figure 6-4: Repair of a structural composite laminate using the cure-in-place concept (Ref. 6-2)

Method No. 6.5: Repair of Structural Laminates Using The Flush Patch Repair Concept

This repair may be used for primary load bearing composite laminate structures. The limits for this repair should be determined by a structures engineer for the structure to which it is to be applied.

NOTE: The laminate should be thoroughly dried prior to the application of the repair material to prevent blistering during cure.

Method of Repair. ***(See Figure 6-5):***

1. Mask off an area which is at least 30 times the thickness of the laminate larger on all sides of the area to be repaired.
2. Clean the masked off area with a trichloroethane dampened clean, white cotton cloth and remove any paint using a method from Chapter 4.
3. Remove the damaged material using a suitable tool. (See Table 2-1).
4. Scribe a circle whose radius is larger than the radius of the area to be removed by 18 times the laminate thickness.
5. Machine away material inside this circle to form a scarf having an L/t ratio of 18/1. (See Figure 6-5, View A.)
6. Cut replacement plies of the same material as the laminate maintaining the same ply orientation. Each successive ply shall be cut larger than the previous ply to match the L/t of the scarf. Two extra plies of the largest two plies shall be cut for use on the bottom side.
7. Cut two pieces of structural film adhesive which are at east one inch larger than the outer edge of the scarf.
8. Place one piece of the film adhesive over the scarf and cut out the bottom of the hole.
9. Assemble the plies of prepreg to fill the hole and build up the top surface. Remember to remove the separator sheets.
10. Place the second sheet of film adhesive over the bottom of the repair and place the two remaining plies over the adhesive withthe largest on the outside. (See Figure 6-5, View B.)
11. Apply a vacuum bag and cure in accordance with the

manufacturer's instructions. If a two sided vacuum bag cure can not be accomplished, a back-up tool may be used to cover the bottom of the repair while the top repair is cured, followed by a second repair and cure of the bottom layers.

12. Feather the outer edges of the cured repair and refinish as required.

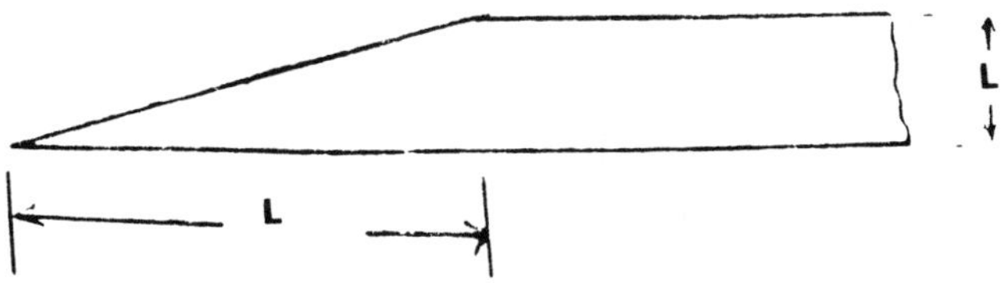

View A

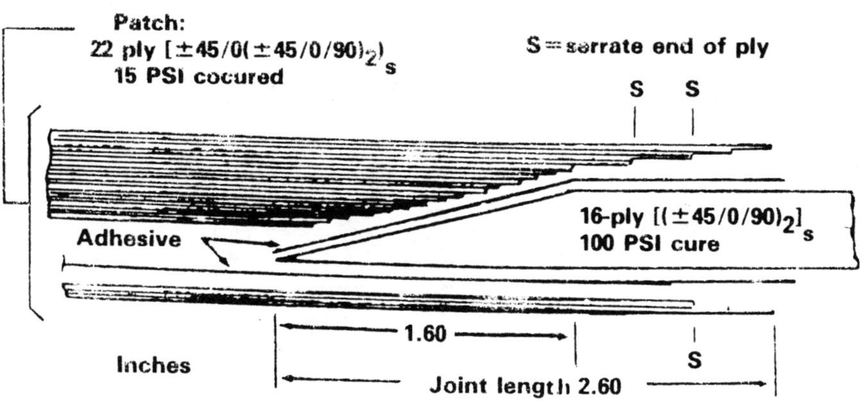

View B

Figure 6-5: Repair of a structural composite laminate using the flush-patch concept. (Ref. 6-2)

REPAIR OF ADVANCED COMPOSITE FACED SANDWICH STRUCTURE

Method No. 6.6: Repair of Composite Faced Structures Containing Dents (No Voids or Cracks).

Method of Repair (See Figure 6-6):

1. Mask off an area approximately two inches larger on all sides than the area to be repaired.
2. Remove paint from the masked off area by an appropriate method. (See Chapter 4.)
3. Clean the surface with trichloroethane using a clean white cotton cloth.
4. Mix the aerodynamic smoother (See Table 2-1) in accordance with the manufacturer's instructions.
5. Apply the mixed aerodynamic smoother with a suitable tool (spatula, tongue depressor, etc.) and smooth to contour.
6. Cure in accordance with the manufacturer's instructions.

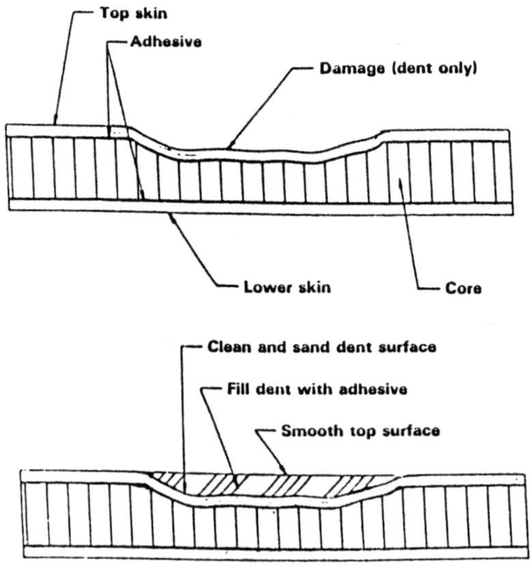

Figure 6-6: Repair of composite-faced structure containing dents (no voids or cracks). (Ref. 6-3)

7. Refinish as necessary to obtain a smooth surface and then paint.

Method No. 6.7: Repair of Composite Faced Structures Containing Tears, Cracks, or Surface Defects.

Method of Repair (See Figure 6-7):

1. Mask off an area approximately three inches larger on all sides than the area to be repaired.
2. Remove paint and/or raised or rough surface by sanding with 100 grit sandpaper followed by 400 grit sandpaper.
3. Clean the surface with a trichloroethane dampened clean white cotton cloth.
4. Cut two pieces of glass cloth so that the first will overlay the damaged area by one inch on all sides and the second

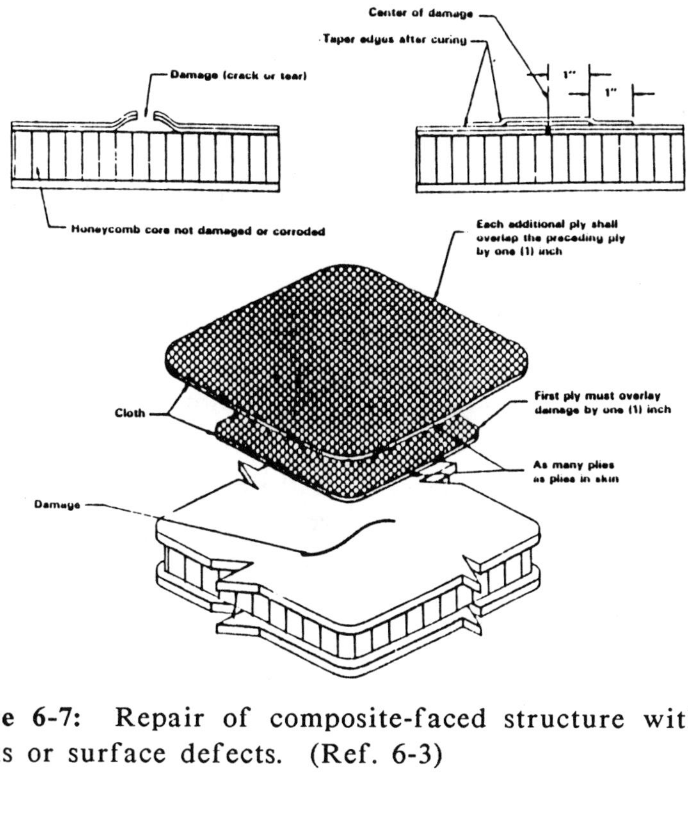

Figure 6-7: Repair of composite-faced structure with tears, cracks or surface defects. (Ref. 6-3)

will be one inch larger on all sides than the first.

5. Mix an appropriate amount of the laminating adhesive (See Table 2-1), and impregnate the pieces of glass cloth.
6. Apply the two layers of impregnated glass cloth so that the first ply is centered over the damaged area and the second ply overlaps the first on all sides by one inch.
7. Cover the repair with a vacuum bag, apply pressure, and cure in accordance with the manufacturer's instructions.
8. Remove the bag and sand to obtain a feather, smooth the surface, paint as required.

Method No. 6.8: Repair of Composite Faced Structures With Minor Damage.

Method of Repair (See Figure 6-8):

1. Mask of an area approximately two inches larger on all sides than the area to be repaired.
2. Remove the paint from the masked area by the appropriate method in Chapter 5.
3. Remove the damaged skin and remove the damaged core using suitable tools such as a hole saw and needle nose pliers. Use care so as not to damage the bottom skin.
4. Flush the cavity and skin with trichloroethane and allow to dry.
5. Mix the repair adhesive (See Table 2-1.) in accordance with the manufacturer's instructions.
6. Fill the hole to over-flush with the adhesive and cover with masking tape.
7. Cure the adhesive in accordance with the manufacturer's instructions.
8. Refinish as necessary to obtain a smooth surface and paint as required.

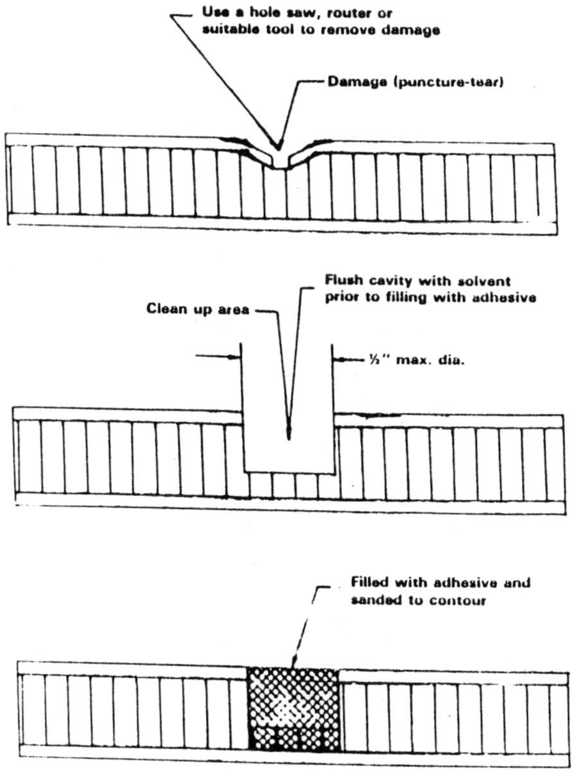

Figure 6-8: Repair of composite-faced structures containing minor core damage. (Ref. 6-3)

Method No. 6.9: Repair of Composite Faced Structures Containing Major Core Damage (Hole in one Skin Only)

Major damage is damage too large to repair with an adhesive plug but requires core replacement.

Method of Repair (See Figure 6-9):

1. Cover the damage with a solvent resistant pressure sensitive tape.
2. Mask off an area approximately four inches larger on all sides than the area to be repaired.

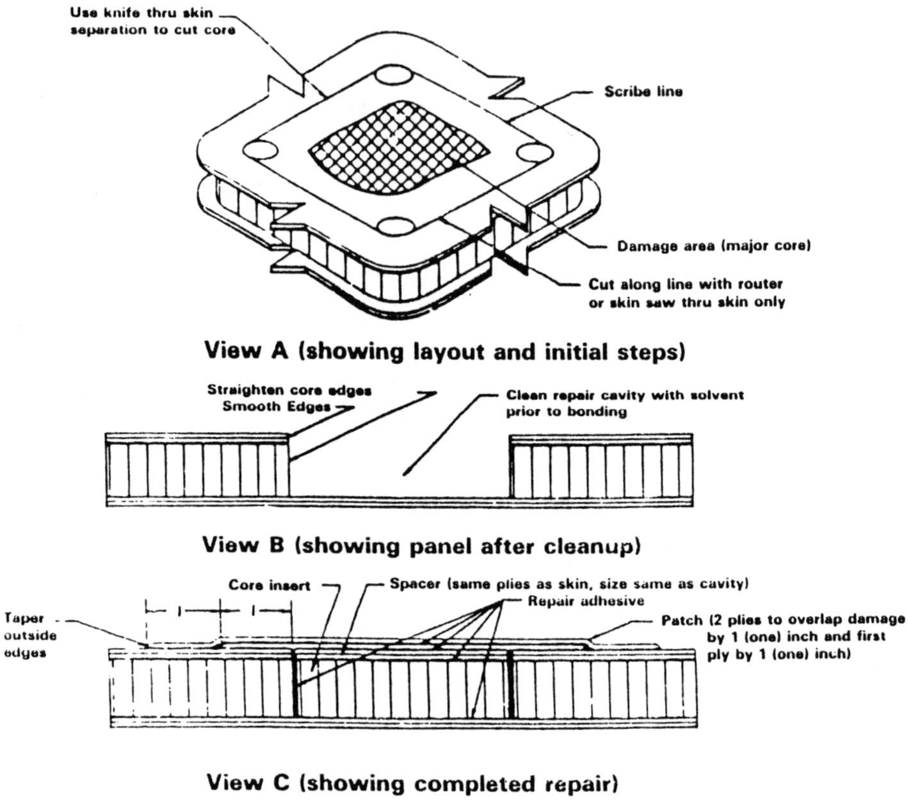

Figure 6-9: Repair of composite-faced structure containing major core damage - hole in one skin. (Ref.6-4)

3. Clean the masked off area with a trichloroethane dampened clean white cotton cloth and remove any paint using a method from Chapter 5.

4. Layout, using a scribe, the skin and core area to be removed. The layout area should have 1/8 inch minimum radius corners, and the scribe marks should not extend past the radii. (See Figure 6-9, View A.)

5. Drill four 1/4 inch minimum diameter holes in the corners of the layout, being careful not to drill into the bottom skin. (See Table 2-1 for special drill bits).

6. Cut the skin between the holes skin using an appropriate tool such as a skin saw, sharp knife, etc. A special tool may be required for cutting some advanced composite skins (See Table 2-1). A depth control must be used on all power tools to protect the core while cutting the skin.
7. Cut through the core with a sharp knife to the opposite skin, following the cut in the skin.
8. Remove the damaged skin.
9. Remove and clean out the affected honeycomb core. (See Figure 6-9, View B.) Needle nose and duck bill pliers have been found useful for this purpose. After removal of the core, the hexagonal impressions should be removed from the adhesive which is adhered to the bottom skin. The adhesive need not be removed if it does not peel from the skin. Rotary files, wire brushes or sanding discs may be used to remove the adhesive if is necessary to do so. NOTE: Excessive pressure on the bottom skin will cause delamination.
10. Smooth all sharp edges in the top skin with a file or suitable grinder.
11. Fabricate a replacement core insert from equal material to fit the removed section. Keep the core ribbon direction the same as the original core.
12. Clean the honeycomb core and the cavity with trichloroethane and dry with clean, dry, oil free compressed air.
13. Mix the repair adhesive (See Table 2-1) in accordance with the manufacturer's instructions.

    NOTE: For large Depot repairs a structural film adhesive and a core splice adhesive may be used. (See Method No. 7.1.)
14. Apply the adhesive to the sides and bottom of the cavity. Place the core insert in position and fill all open areas in the bondline with adhesive.
15. Apply vacuum bag pressure and cure in accordance with the adhesive manufacturer's instructions. Remove the bag.

16. Cut a glass cloth insert the same size as the core insert and of the same number of plies as the original skin.
17. Mix the laminating adhesive (See Table 2-1) in accordance with the manufacturer's instructions.
18. Impregnate the plies of glass cloth and position them over the core insert. If possible attempt to maintain the original orientation.
19. Add two more plies of impregnated glass cloth so that the first will overlap the repair by one inch on all sides. The second should overlap the first by one inch on all sides. (See Figure 6-9, View C.)
20. Cover the repair with a release film material and then with an aluminum plate (1/8 inch thick minimum) which is approximately 1/4 inch larger than the last ply, apply a vacuum bag and cure in accordance with the adhesive manufacturer's instructions.
21. Remove the bag and sand the repair to obtain a smooth surface and feathered edges; paint as required.

Method No. 6.10: Repair of Composite Faced Structures Containing Major Core Damage (Hole in Both Skins)

Method of Repair (See Figure 6-10):

1. Cover the damage with a solvent resistant pressure sensitive tape.
2. Mask off an area approximately four inches larger on all sides than the area to be repaired.
3. Clean the masked off area with a trichlorethane dampened clean white cotton cloth and remove any paint using an appropriate method (See Chapter 4).
4. Layout the skin and core area to be removed, using a scribe. The layout area should have 1/8 inch minimum radius corners and the scribe lines should not extend past the radii.

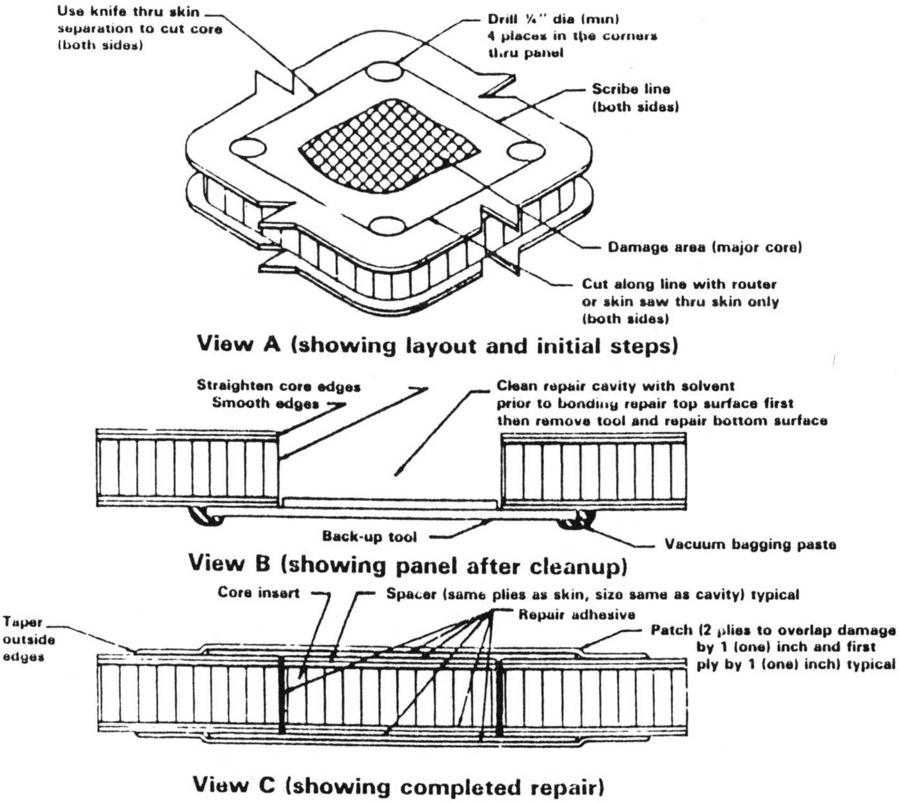

Figure 6-10: Repair of composite-faced structure containing major damage - Hole in both skins. (Ref.6-4)

5. Drill four 1/4 inch diameter holes in the corners through the panel. (See Table 2-1 for special drill bits).
6. Cut the skin between the holes, on both sides with an appropriate tool.
7. Cut through the core using a sharp knife and remove the skins and the core.
8. Prepare a metal spacer which is the same thickness as the skin and attach it to a larger back plate with a minimum of two rivets. (See Figure 6-11.) Apply a non-silicone release agent to the surface of the back-up plate.

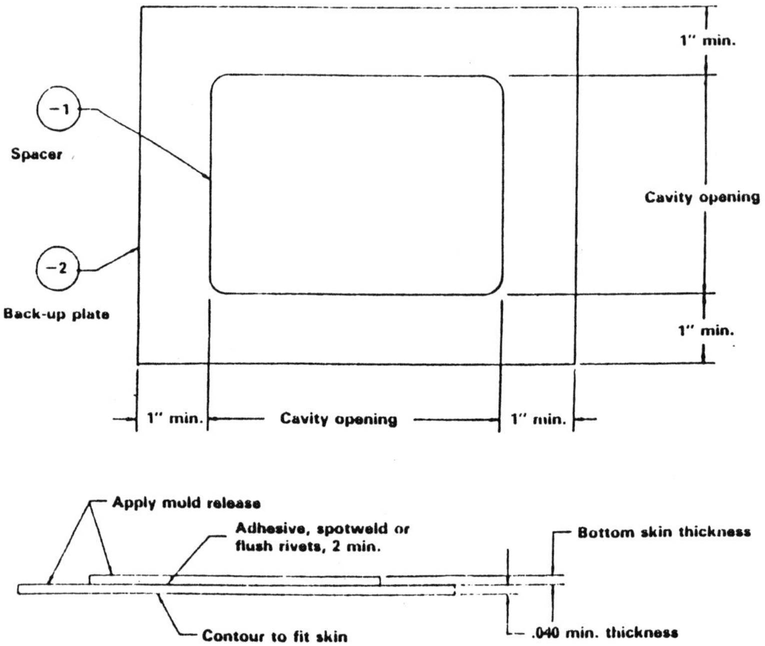

Figure 6-11: Back-up plate for honeycomb structure repair (Ref.6-4)

9. Fabricate a replacement core insert from equal material to fit into the removed section. Keep the core ribbon direction the same as in the original core. Straighten all the core edges in the repair area and the insert.

10. Clean the cavity and the insert with a trichloroethane spray and dry with clean, dry, oil free compressed air.

11. Position the back-up plate into the hole. Seal the outside edges with vacuum bagging putty. (See Figure 6-10, View B)

12. Mix the repair adhesive (See Table 2-1) in accordance with the manufacturer's instructions.

    NOTE: For large Depot repairs a structural film adhesive and core splice adhesive may be used. (See Method No. 7.2.)

13. Apply the adhesive to the sides of the cavity and the core insert. Place the insert into the cavity.
14. Cut glass cloth plies the same size as the core insert and of the same number of plies as the original skin.
15. Mix the laminating adhesive (See Table 2-1) in accordance with the manufacturer's instructions.
16. Impregnate the glass cloth with the laminating adhesive and position them over the core insert.
17. Add two more plies of glass cloth impregnated with the laminating adhesive. The first ply shall overlap the insert by one inch on all sides and the second ply shall overlap the first by one inch on all sides. (See Figure 6-10, View C.)
18. Cover with a sheet of release film and place an aluminum plate (1/8 inch thick minimum) which is approximately 1/4 inch larger than the last ply.
19. Apply vacuum bag pressure and cure in accordance with the adhesive manufacturer's instructions.
20. Remove the bag and the back-up plate and repeat steps 14 through 19.
21. Sand the repair patch on both sides to obtain smooth and feather edges. Paint as required.

Method No. 6.11: Repair of Composite Faced Structures Having an Inaccessible (Blind-Side) Back Skin (Ref.6.4)

When the back skin of a structure is not accessible and the part can not be removed the structure can be repaired by the following procedure. (See Figure 6-12)

Method of Repair:

1. Prepare the structure for repair in accordance with the Method of Repair given in Method No. 6.10 following Steps 1 through 3.

2. Cut out the damaged area, using a 2", 3", or 4.5" diameter hole saw.

3. Sand and clean all bondable surfaces in the repair area.

4. Select The blind-side repair materials as described in Figure 6-13 (Atacs repair kit number 5900 or equivalent), sized to fit the hole cut in Step 2.

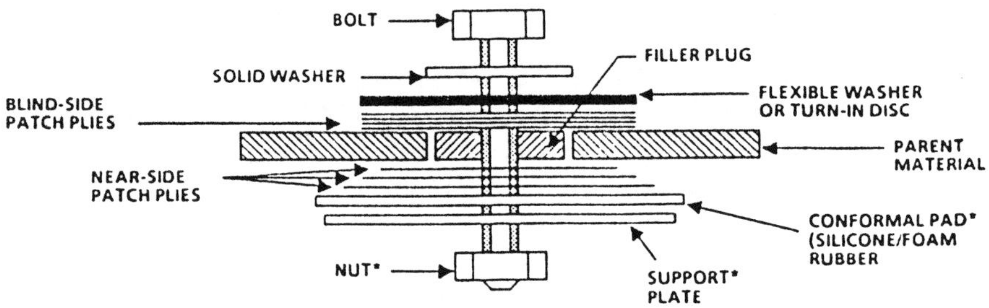

Figure 6-12: Patch assembly for blind-side repair.

5. Select the the core material to fit the repair hole, the thickness should be equal to the thickness of the original panel (See Figure 6-16.)

6. Mix a repair adhesive (See Table 2-1).

7. Impregnate fabric plies as follows: Place the blind-side repair material on a sheet of polyethylene or equivalent film. Pour the mixed resin onto the repair material. Fold the polyethylene film over the repair material and work the resin thoroughly into the repair material checking the reverse side periodically for complete wet out. Be sure the plies are fully saturated and the excess resin is removed by working for the center outward.

8. Assemble the blind-side patch material onto the bolt in the following order: Solid washer, turn-in disc or flexible washer (use turn-in disc for thin structures, and flexible

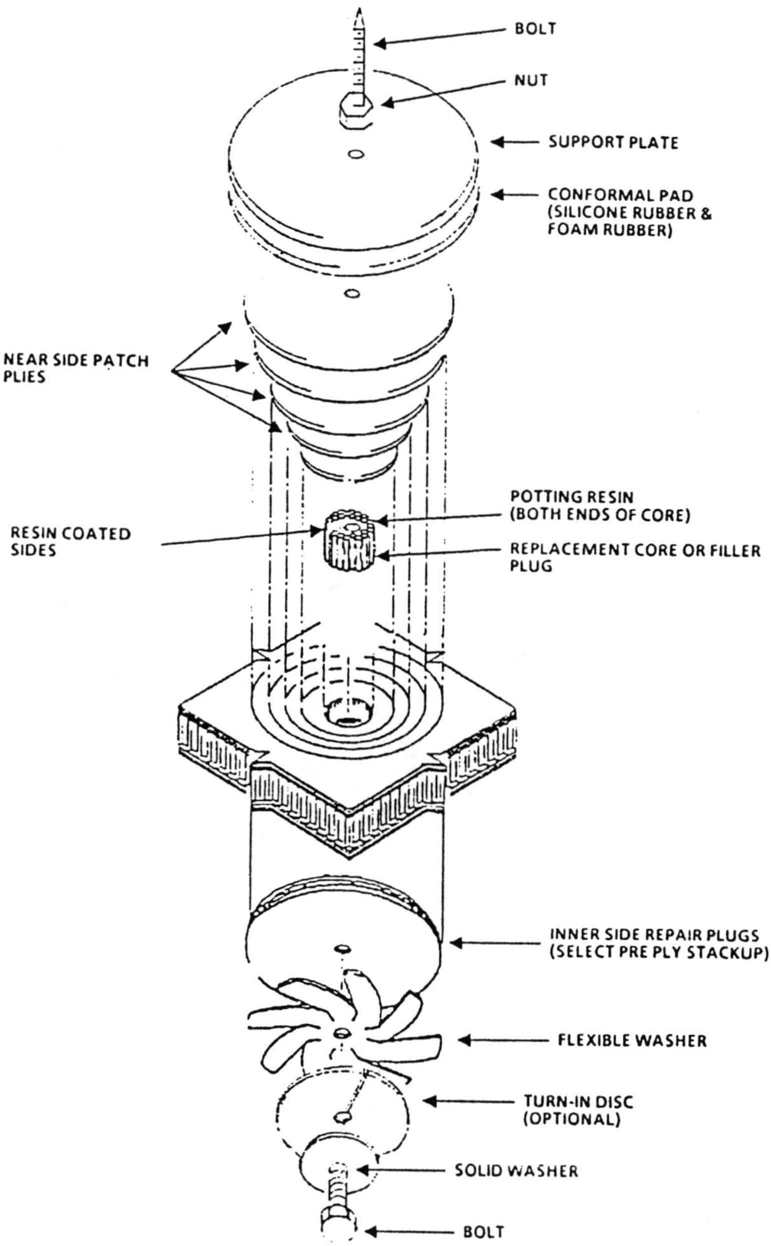

Figure 6-13: Repair of damage to both skins with access limited to one side.

washer for sandwich and thick structures), repair plies and insertion tool.

9. Push the blind-side repair patch assembly through the hole using the insertion tool (See Figure 6-14 and 6-15).
10. Pull the patch up against the backside of the hole and remove the insertion tool.
11. Apply repair adhesive to the edge of the core replacement plug and insert the plug into the cavity.

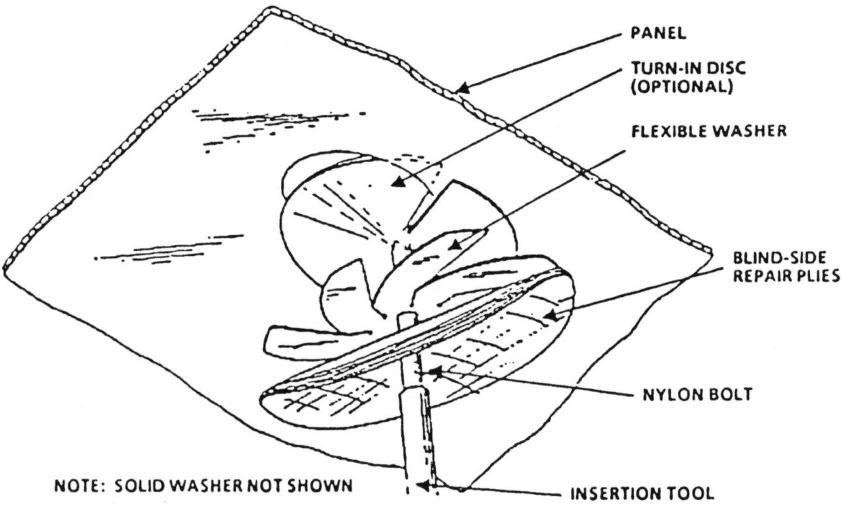

Figure 6-14: Installation using turn-in disc.

12. Slide the near-side repair material over the bolt shank and against the near side surface.
13. Slide the conformal pad, and the support plate over the shank and against the near side surface. NOTE: Be certain that a parting film or material has been used between the conformal pad and the near side patch unless it is to become a part of the repair. (A silicone rubber conformal pad can be used as a parting material.)

14. Install and tighten the nut, 5-10 in-lbs. (if using a wing nut, cinch snugly.)
15. Cure in accordance with the adhesive manufacturer's instructions.
16. Remove the nut, the support plate and the conformal pad. (Turning the nut will shear the nylon bolt off near the surface.)
17. Shave the bolt flush with the surface (See Figure 6-16.)
18. Smooth and finish to match the existing surface.

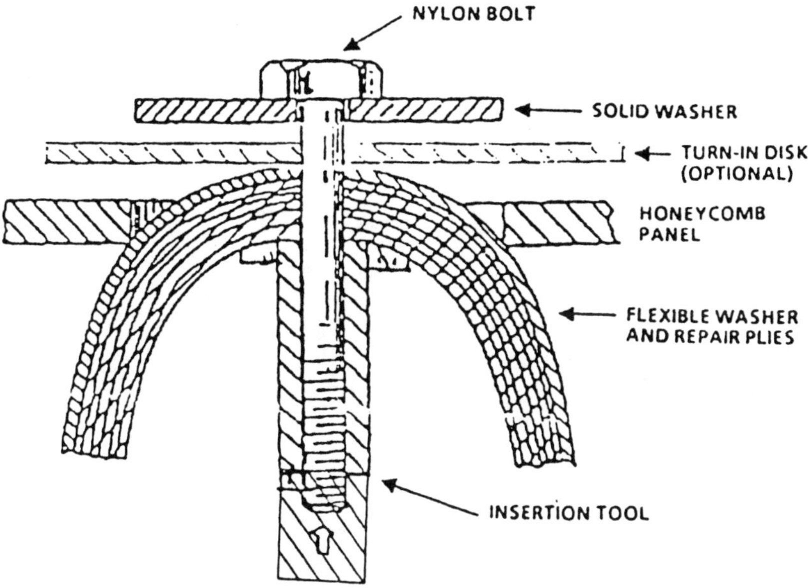

Figure 6-15: Installation using flexible washer.

Method No. 6.12: Repair of Composite Faced Structures Containing Edge Voids

Edge voids must not be considered negligible damage and any panel that has lost its edge seal shall be resealed.

Method of Repair (See Figure 6-17):

1. Using a drill bit the same size as the rivet shank, drill through the center of the rivet head to the panel surface.
2. Using drift punch, pop off the rivet head.
3. Lightly drive out the rivet shank using a drift punch. NOTE: For blind lick bulb rivets, a rivet shaver or grinder may be used in lieu of Steps 2 and 3.
4. Mix the repair adhesive in accordance with the manufacturer's instructions.

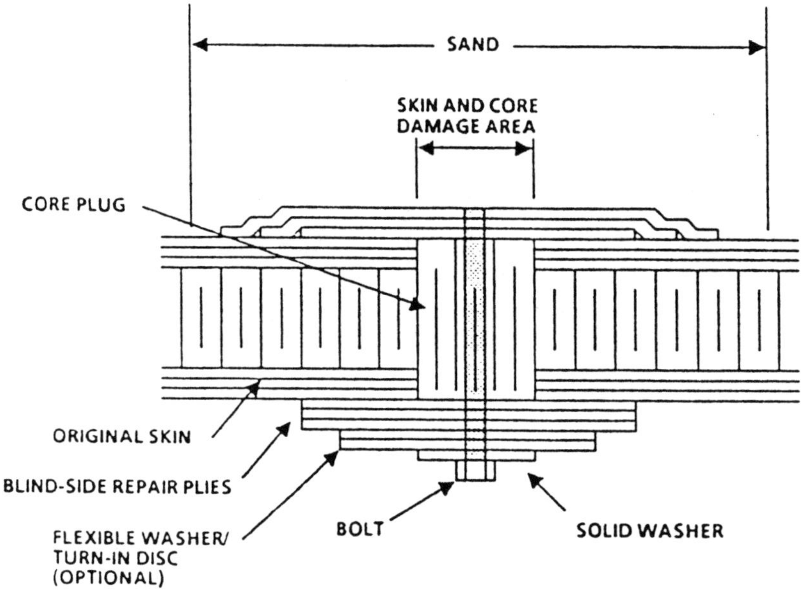

Figure 6-16: Completed blind-side repair.

5. Using an injection needle (See Table 2-1) and syringe inject the adhesive directly into the void from a separated edge.

6. Immediately apply pressure to the void to allow the excess adhesive to flow from the rivet holes.
7. Cure the adhesive under pressure, in accordance with the manufacturer's instructions.
8. Seal the edges with sealing compound and allow the sealant to cure per instructions.

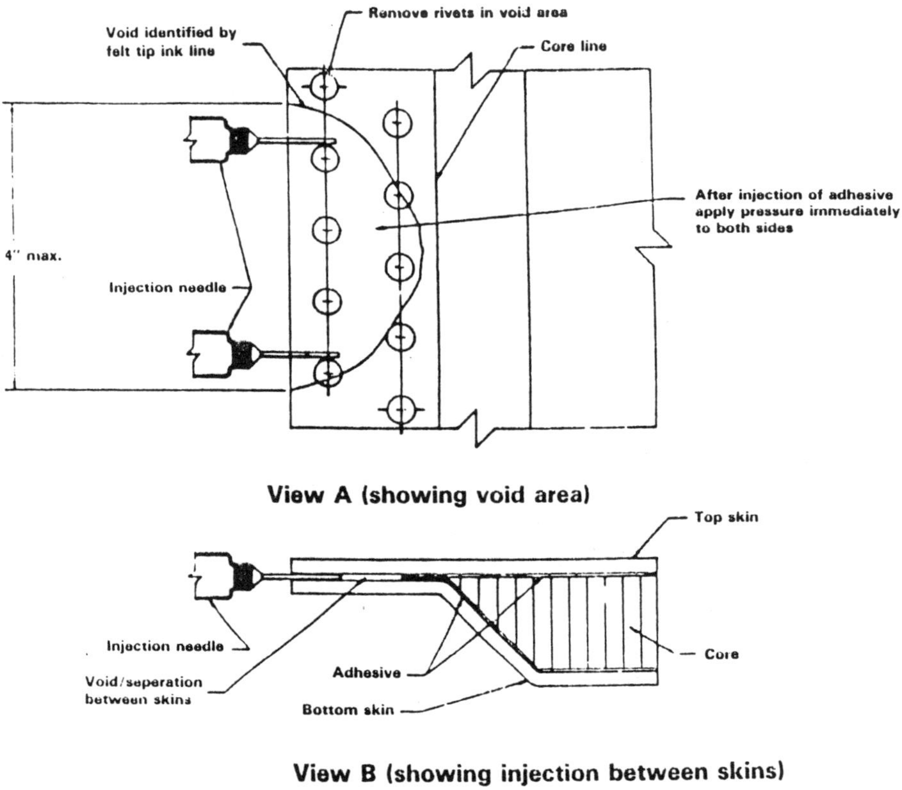

Figure 6-17: Repair of composite-faced structure containing edge voids. (Ref.6-3)

REFERENCES

6-1. Knauss, J.F. and Stone, R.H., "Demonstration of Repairability and Repair Quality on Graphite/Epoxy Structural

Subelements", 27th National SAMPE Symposium, May, (1982).

6-2. Stone, R.H., "Development of Repair Procedures for Graphite/Epoxy Structures on Commercial Transports", SAMPE Monograph No.1, Edited by Henry Brown, SAMPE, (1985).

6-3. Wegman,R.F., "Repair of Damage to Secondary Aircraft Structures of Advanced Composite Materials", PLASTEC Report No. R-51, October, (1983).

6-4. Atacs Repair Procedure 5900. Atacs Materials Inc, Seattle WA.

7

LARGE AREA DEPOT REPAIR OF COMPOSITE FACED STRUCTURES

There may be limitations as to the size of repair that can be made by a particular maintenance facility. These are, to a large extent, determined by the availability of special equipment such as autoclaves, bonding tools and alignment fixtures. For these types of repairs, the structure often must be returned to a factory or depot where the necessary facilities and personnel are available.

DAMAGE REMOVAL METHODS

If the damaged material is properly removed the subsequent repair operation can be made more easily. Care must be taken so as not to extend the damage during the clean-out operation. Proper tools should be used to cleanly cut the skins and the core. Some adhesives, especially those used in high temperature applications may peel easily. Special care must be taken to prevent the peeling of the skin from causing delamination beyond the repair boundary. Paint strippers should never be used to remove paint from a composite structure.

REMOVAL OF SKIN MATERIALS

Numerous types of mechanical equipment are available to cut, slit, rout or saw composite skins. The high speed router is convenient and versatile for this purpose. The depth of cut can be controlled and the outline of the removal area can be defined by using a template. The slotting saw is also a convenient tool for the removal of skins. The composite skin is cut into strips which can then be peeled off in small pieces.

REMOVAL OF CORE MATERIAL

The removal of light density core may often be accomplished using a sharp knife or putty knife. Duckbill pliers are useful. The core should be removed down to the adhesive bondline. If adhered to the structure the adhesive need not be removed. Heavier density cores may require machining. An abrasive disc with a portable high-speed motor may be used. Following core removal, the adhesive surface may be smoothed using a portable high speed motor and a sanding disc. Care should be taken not to apply too much pressure during the sanding operation or further delamination may occur. After the smoothing procedure use a vacuum cleaner to remove dust and debris. Dry wipe with a clean cheesecloth or cloth.

IN-PROCESS QUALITY ASSURANCE

Post repair inspection techniques such as nondestructive testing are not sufficient to ensure that a quality repair has been made. Properties such as high bond strengths, reliability and durability can only be accomplished through careful step-by-step processing and the use of in-process controls. Special consideration must be given to:

1. Using the approved processes.
2. Using only approved materials.

3. Prefit of detail parts before bonding.
4. Proper handling and application of bonding materials.
5. Adherence to the recommended cure cycles.

METHODS OF REPAIR

The methods of repair for the following defects at a depot or factory are the same as for field repairs. The repair of these defects are limited by size of the defect. When the size of the defect exceeds the limits set for the repair there could be structural damage that is not apparent. Therefore, these larger defects should be treated as major damage and handled accordingly:

1. Repair of composite faced structures containing dents.
2. Repair of composite faced structures with tears, cracks or surface defects.
3. Repair of composite faced structure with minor core damage.

Method No. 7.1: Repair of Composite Faced Structures Containing Major Core Damage (Hole in one skin only)

Major damage is damage too large to repair with an adhesive plug but requires core replacement.

Method of Repair:

1. Cover the damage with a solvent resistant pressure sensitive tape.
2. Mask off an area approximately four inches larger on all sides than the area to be repaired.
3. Clean the masked off area with a trichloroethane dampened clean white cotton cloth and remove any paint using a method in Chapter 4.

4. Layout, using a scribe, the skin and core area to be removed. The layout area should have 1/8 inch minimum radius corners, and the scribe marks should not extend past the radii.

5. Drill four 1/4 inch minimum diameter holes in the corners of the layout, being careful not to drill into the bottom skin. (See Table 2-1 for special drill bits).

6. Cut the skin between the holes using an appropriate tool such as a skin saw, sharp knife, etc. A special tool may be required for cutting some advanced composite skins (See Table 2-1). A depth control must be used on all power tools to protect the core while cutting the skin.

7. Cut through the core with a sharp knife to the opposite skin, following the cut in the skin.

8. Remove the damaged skin.

9. Remove and clean out the affected honeycomb core. Needle nose and duck bill pliers have been found useful for this purpose. After removal of the core, the hexagonal impressions should be removed from the adhesive which is adhered to the bottom skin. The adhesive need not be removed if it does not peel from the skin. Rotary files, wire brushes or sanding discs may be used to remove the adhesive if necessary.

   NOTE: Excessive pressure on the bottom skin will cause delamination.

10. Smooth all sharp edges in the top skin with a file or suitable grinder.

11. Fabricate a replacement core insert from equal material to fit the removed section. Keep the core ribbon direction the same as the original core.

12. Clean the honeycomb core and the cavity with trichloroethane and dry with clean, dry, oil free compressed air.

13. Cut a section of the structural adhesive (See Table 2-1) to fit the bottom of the cavity.

    NOTE: Structural adhesives must be stored at temperatures

below zero degrees F. They should be removed from storage at least 12 hours before they are used to allow the adhesive to come to room temperature before removal from their protective packaging.

14. Apply the adhesive to the sides and bottom of the cavity. Place the core insert in position and fill all open areas in the bondline with adhesive.

15. Cut a strip of core splice adhesive (See Table 2-1) to fit the sides of the core adhesive and apply the adhesive by pressing it to the edge of the core (See Figure 7.1). Remove the other separator film. Place the insert into the cavity.

16. Apply vacuum bag pressure and cure in accordance with the adhesive manufacturer's instructions. Remove the bag.

17. Cut a glass cloth insert the same size as the core insert and of the same number of plies as the original skin.

    NOTE: Prepreg material may be used to replace skins if prepreg was used in the original fabrication. In this case Steps 18, 19 and 20 would be omitted. If prepreg material is used, the orientation should be the same as in the original skin.

18. Mix the laminating adhesive (See Table 2-1) in accordance with the manufacturer's instructions.

19. Impregnate the plies of glass cloth and position them over the core insert. If possible, maintain the original orientation.

20. Add two more plies of impregnated glass cloth so that the first will overlap the repair by one inch on all sides. The second should overlap the first by one inch on all sides.

21. Cover the repair with a release film material and then with an aluminum plate (1/8 inch thick minimum) which is approximately 1/4 inch larger than the last ply, apply a vacuum bag and cure in accordance with the adhesive manufacturer's instructions.

22. Remove the bag and sand the repair to obtain a smooth surface and feathered edges. Paint as required.

Method No. 7.2: Repair of Composite Faced Structures Containing Major Core Damage (Hole in both skins)

Method of Repair:

1. Cover the damage with a solvent resistant pressure sensitive tape.
2. Mask off an area approximately four inches larger on all sides than the area to be repaired.
3. Clean the masked off area with a trichlorethane dampened clean white cotton cloth and remove any paint (See Chapter 4).
4. Layout the skin and core area to be removed, using a scribe. The layout area should have 1/8 inch minimum radius corners and the scribe lines should not extend past the radii.
5. Drill four 1/4 inch diameter holes in the corners through the panel. (See Table 2-1 for special drill bits).
6. Cut the skin between the holes, on both sides with an appropriate tool.
7. Cut through the core using a sharp knife and remove the skins and the core.
8. Prepare a metal spacer which is the same thickness as the skin and attach it to a larger back plate with a minimum of two rivets. Apply a non- silicone release agent to the surface of the back-up plate.
9. Fabricate a replacement core insert from equal material to fit into the removed section. Keep the core ribbon direction the same as in the original core. Straighten all the core edges in the repair area and the insert.
10. Clean the cavity and the insert with a trichloroethane spray and dry with clean, dry, oil free compressed air.

11. Position the back-up plate into the hole. Seal the outside edges with vacuum bagging putty.

12. Cut a section of the structural adhesive (See Table 2-1) to fit the bottom of the cavity.

    NOTE: Structural adhesives must be stored at temperatures below zero degrees F. They should be removed from storage at least 12 hours before they are used to allow the adhesive to come to room temperature before removal from their protective packaging.

13. Cut a strip of core splice adhesive (See Table 2-1) to fit the sides of the core adhesive and apply the adhesive by pressing it to the edge of the core, remove the other separator film. Place the insert into the cavity. Cure under pressure in accordance with the adhesive manufacturer's instruction.

14. Cut glass cloth plies the same size as the core insert and of the same number of plies as the original skin.

    NOTE: Prepreg material may be used to replace skins if prepreg was used in the original fabrication. In this case Steps 15, 16 and 17 would be omitted. If prepreg is used the orientation should be the same as in the original skin.

15. Mix the laminating adhesive (See Table 2-1) in accordance with the manufacturer's instructions.

16. Impregnate the glass cloth plies with the laminating adhesive and position them over the core insert.

17. Add two more plies of glass cloth impregnated with the laminating adhesive. The first ply will overlap the insert by one inch on all sides and the second ply should overlap the first by one inch on all sides.

18. Cover with a sheet of release film and place an aluminum plate (1/8 inch thick minimum) which is approximately 1/4 inch larger than the last ply.

19. Apply vacuum bag pressure and cure in accordance with the adhesive manufacturer's instructions.

20. Remove the bag and the back-up plate and repeat Steps 14 through 19.

21. Sand the repair patch on both sides to obtain smooth and feathered edges. Paint as required.

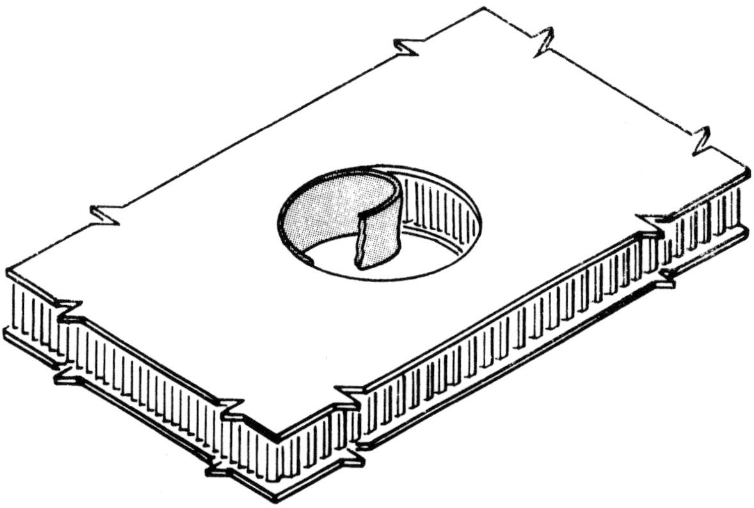

Figure 7-1: Application of core splice adhesive to the edge of the cavity. (Ref. 7-1)

DEPOT REBUILD OF COMPOSITE FACED STRUCTURES

Repair involves leaving the bulk of the structure intact and fixing areas within certain established limits. When the limits for repair have been exceeded there are three basic options:

1. Replacement of the structure with a new part from stock is available,
2. Rebuild the structure, or
3. Fabricate a new part.

Rebuilding a composite faced structure entails major reconstruction in which an entire skin and most of the core material may be replaced. Some critical parts of the original structure are reused. Rebuilding involves the removal of the damaged skin, removal of the necessary core, preparation of new core, the fabrication of a new skin and the assembly and rebonding of the component parts using heat curing structural adhesives.

Method No. 7.3: Rebuild of Composite Faced Structures

This method involves the replacement of a major portion of the structure with new materials that are equal to the original design materials and structural adhesives

Method of Rebuild:

1. Remove the damaged or delaminated skin taking care not to damage any undamaged core. Numerous types of mechanical equipment are available to cut , slit, rout or saw the composite skin.

   Equipment commonly used includes the high speed router, skin saw, or slotting saw. A sharpened putty knife may be used to lift the edge of the skin, after which it may be gripped with a pair of pliers, vise grips or a slotted Tee-Bar tool. When attempting to remove a skin, be careful to pull slowly and to be aware that the composite skin may be brittle and easily broken.

   If the adhesive has a high peel strength dry ice may be applied to the skin surface. This tends to embrittle the adhesive and lower the peel strength.

   NOTE: Dry ice may also embrittle the composite skin. Dry ice should be handled with insulated gloves and safety glasses should be worn. Serious frostbite or burn may result if dry ice comes in contact with the skin or eyes. In case of dry ice burn obtain medical assistance immediately. Provide ventilation for carbon dioxide vapors.

   In some cases the use of heat may be effective in

helping remove a skin. Heat may lower the strength of the adhesive. Heat may be applied by a heating blanket or heat gun. Again care must be taken to prevent skin burns.

2. Remove damaged core using a suitable tool which will not damage the remaining core. A sharpened putty knife, router or abrasive disc on a high speed motor may be used. The core is removed down to the adhesive layer on the bottom skin which need not be removed if it is bonded to the skin.

3. If fuel or oil is detected, determine if any damage to the remaining structure has occurred. If the part is determined salvageable the assembly should be cleaned with trichloroethane and dried. If moisture is detected it should be removed by the application of heat from heat lamps or other source. The temperature should not exceed 180 degrees F.

4. Remove core chips and debris with a vacuum cleaner or other suitable means. Clean the exposed surfaces with a trichloroethane dampened clean white cotton cloth to remove any traces of sanding dust.

5. Prefit the replacement core using the same material and density as the removed core. Replacement honeycomb core shall be fabricated so that the ribbon direction matches the ribbon direction of the original core If precut core is not available, it may be cut with a band saw or other suitable means. Special tools and equipment may be required when contoured honeycomb shapes are involved. Some of this equipment is described in MIL-HDBK-337 "Military Standardization Handbook: Adhesive Bonded Aerospace Repair", Section 7.

   Thickness should be maintained within 0.010 inches. Splices in the core shall be butt joints and should be made with a core splice adhesive.

6. Fabricate and prefit any required internal parts using the same material as the original. Prepare for bonding as required.

7. Fabricate and prefit a new skin of the same material and thickness as the original. If possible used precured skins with peel plies on the bonding surface. These can be cut to size and have the peel ply removed prior to bonding. If precured skins are not available, prepreg laminate materials may be used. Prepreg laminates may or may not require the use of additional adhesive depending upon the resin content of the prepreg. If necessary the skin may be fabricated by the use of wet layup techniques. These techniques require the use of special tooling.
8. Prefit all parts to ensure a good fit.
9. Remove the adhesive to be used from cold storage and allow it to equilibrate to room temperature before opening the container.
10. Cut the core splice adhesive to fit any joints that will have to be made in the core.
11. Cut the film adhesive to fit the detail parts to be assembled and to fit the skin.
12. Remove the separator sheet from one side of the adhesive and apply the adhesive to the faying surface of the parts to be assembled, being careful to prevent air entrapment. For large parts, lay the adhesive down from the center of the part and work out toward the edges. Leave the separator sheet on the side of the adhesive exposed to the atmosphere as a temporary protective cover. If the adhesive film is used in cut sections there must be no gaps between pieces of adhesive, overlapping of the adhesive film is not recommended unless required for tolerance build up or in beveled edges.
13. Install the core replacement in to the panel assembly. Apply core splice adhesive between butt joints in the core. Core splice adhesive should also be used between core and edge members if applicable.
14. Install any internal parts after removal of the separator sheet from the adhesive. Care should be taken in the

assembly of parts in the curing fixture or jig to assure that the adhesive is not disturbed.

15. Remove the separator sheet from the adhesive on the skin and carefully place the skin on to the panel.
16. Place the part in a vacuum bag and cure in an autoclave in accordance with the manufacturer's instructions. If prepreg materials are used to fabricate the skin and it is to be co-cured with the assembly, the cure temperatures of the prepreg and the adhesives used must be compatible.

FABRICATION OF NEW STRUCTURES

When a structure can not be repaired, rebuilt, or replaced from stock a new structure may be fabricated. This is done in accordance with the original drawings and specification and their latest revisions. The materials of fabrication shall be the same as in the original structure.

REFERENCES

7-1. MIL-HDBK-337, Military Standardization Handbook-Adhesive Bonded Aerospace Structure Repair, 1 December, (1982).

8

FIELD LEVEL REPAIR OF METAL FACED HONEYCOMB STRUCTURES

The intent of this chapter is to present standardized repair technology for metal faced sandwich structures at the field level. These methods may be accomplished with minimal equipment and with two part room temperature curing paste adhesives. The limitations on the size of each specific damage to be repaired will be established by the specific structures manual or by the authorized engineering authority.

Method No. 8.1: Repair of Metal Faced Structures Containing Dents (No Cracks or Voids)

Method of Repair:

1. Mask off an area approximately two inches larger on all sides than the area to be repaired.
2. Remove the paint from the masked off area using the method described in Chapter 4 and then sand with 400 grit sandpaper.
3. Mix the aerodynamic smoother (See Table 2-1) in accordance with the manufacturer's instructions.
4. Apply the mixed aerodynamic smoother with a suitable

tool (spatula, tongue depressor, etc.) and smooth to contour. (Figure 8-1)

5. Cure in accordance with the manufacturer's instructions.
6. Refinish as necessary to obtain a smooth surface, prime and paint as required.

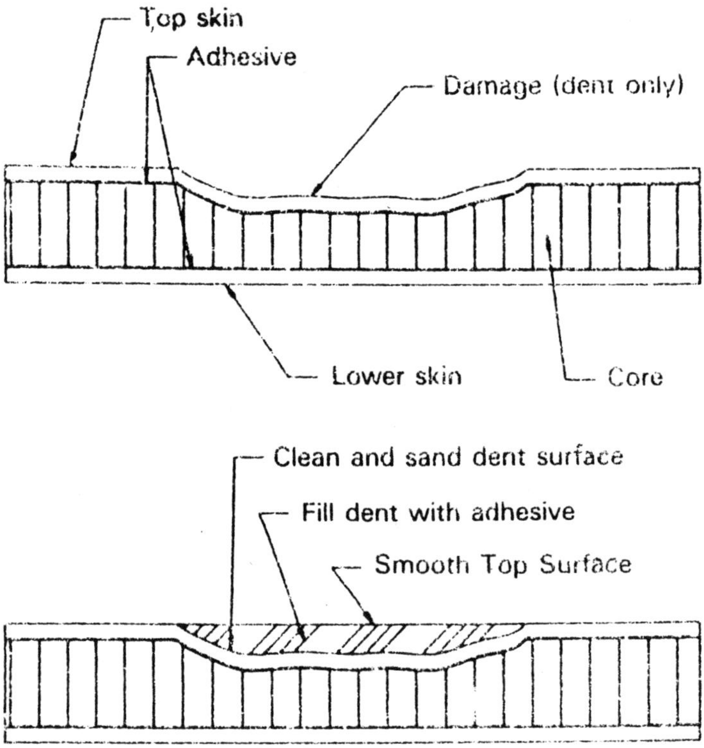

Figure 8-1: Repair of metal-faced structure containing dents (no cracks or voids).

Method No. 8.2: Repair of Metal-Faced Structures Containing a Void Under a Dent (Backside of Structure Accessible)

Sharp dents and creases should be treated as fractures and repaired as such.

Method of Repair:

1. Mask off an area approximately two inches larger on all sides then the area to be repaired.
2. Remove the paint from the marked off area using a method described in chapter 4 then sand with 400 grit sandpaper.
3. Using a No. 40 drill, drill a hole from the backside of the structure at each end of the void, being careful not to drill through the front face.
4. Mix the injection type adhesive (See Table 2-1) in accordance with the manufacturer's instructions.
5. Inject the adhesive into one hole until it flows out the opposite hole. CAUTION: Inject slowly so as not to build up excessive pressure. (See Figure 8-2.)
6. Cover the holes with a piece of masking tape.
7. Apply pressure to the voided skin and cure the adhesive in accordance with the manufacturer's instructions.
8. Mix aerodynamic smoother in accordance with the manufacturer's instructions.
9. Apply the mixed aerodynamic smoother with a suitable tool and smooth to contour.
10. Cure in accordance with the manufacturer's instructions.
11. Refinish as necessary to obtain a smooth surface, prime and paint as required.

Method No. 8.3: Repair of Metal Faced Structures Containing a Void Under a Dent (Backside of Structure not Accessible)

Sharp dents and creases should be treated as fractures.

Method of Repair:

1. Mask off an area approximately two inches larger on all sides than the area to be repaired.
2. Remove the paint from the metal in this area using an appppropriate method from Chapter 4, then sand with 400 grit sandpaper.
3. Using a No. 40 drill, drill a hole at each end of the void and at the center. Care must be taken to minimize penetration of the adhesive film on the core.
4. Mix an injection type adhesive (See Table 2-1) in accordance with the manufacturer's instructions.
5. Inject the adhesive into the outer two holes until it flows out the center hole. (Figure 8-3) NOTE: If an excessive amount of adhesive is required the film adhesive has

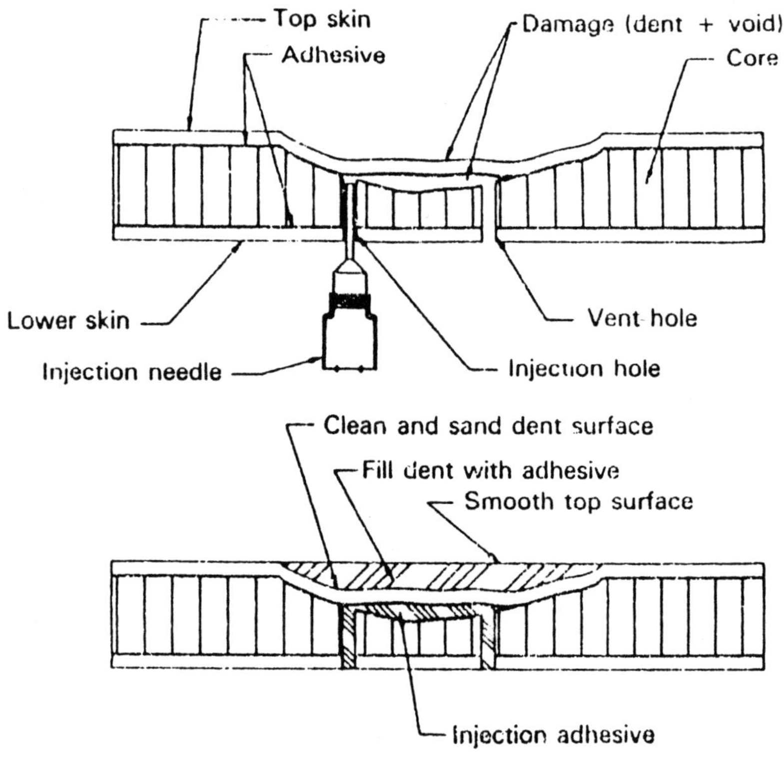

Figure 8-2: Repair of metal-faced structure containing void under dent (backside accessible).

probably failed from the core and the core is being filled with adhesive. Consideration should be given to repairing this damage as a fracture.

6. Cover the holes with a piece of masking tape, apply pressure to the voided skin.
7. Cure in accordance with the manufacturer's instructions.
8. Mix the aerodynamic smoother in accordance with the manufacturer's instructions.
9. Apply the mixed aerodynamic smoother with a suitable tool and smooth to contour.
10. Cure in accordance with the manufacturer's instructions.
11. Refinish as necessary to obtain a smooth surface, prime and paint as required.

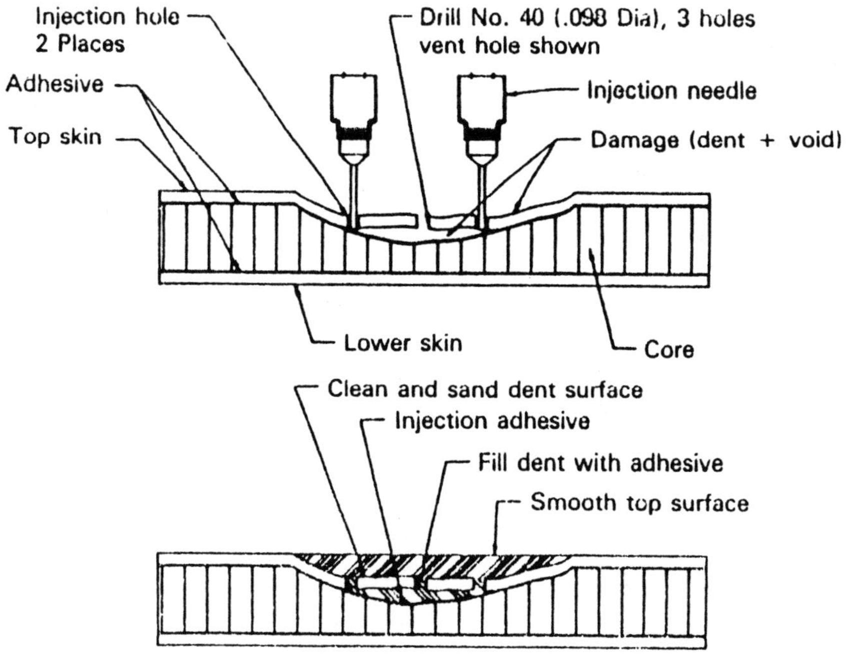

Figure 8-3: Repair of metal-faced structure containing void under dent (backside not accessible).

Method No. 8.4: Repair of Metal Faced Structures Containing Dents With a Hairline Crack in the Dent.

Sharp dents and creases should be treated as fractures.

Method of Repair:

1. Mask off an area approximately two inches larger on all sides than the area to be repaired.
2. Remove the paint from the masked off area by carefully sanding the surface with 100 grit sandpaper followed by a light sanding with 400 grit sandpaper.
3. Remove any sanding dust by using a soft bristle brush.
4. Stop-drill the ends of the cracks using a 1/8 inch drill. (See Figure 8-4.)
5. Mix an aerodynamic smoother (See Table 2-1) in accordance with the manufacturer's instructions.
6. Apply the adhesive with a suitable tool and smooth to contour.
7. Cure in accordance with the manufacturer's instructions.
8. Refinish as necessary to obtain a smooth surface, prime and paint as required.

Method No. 8.5: Repair of Metal Faced Structures Containing Minor Core Damage

Method of Repair:

1. Mask off an area approximately two inches larger on all sides than the area to be repaired.
2. Remove the damaged skin and clean out the damaged core using a suitable tool (hole saw and needle nose pliers). (See Figure 8-5.)
3. Flush out the cavity with trichloroethane and allow to dry.

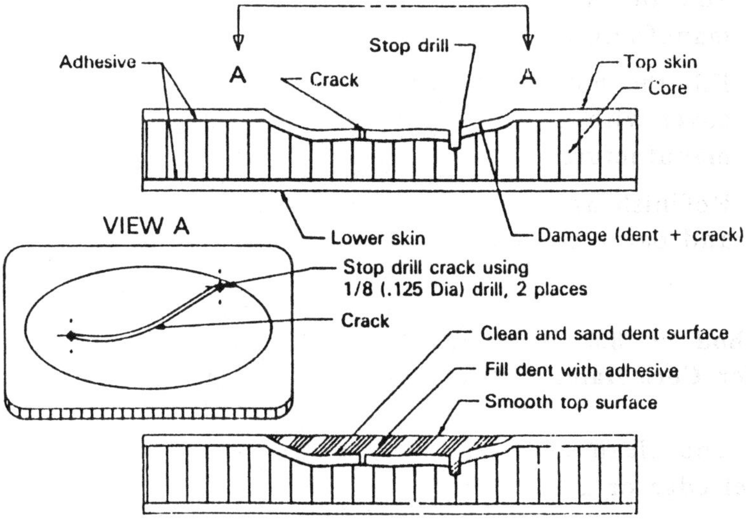

Figure 8-4: Repair of metal-faced structure containing dent with hairline crack in dent.

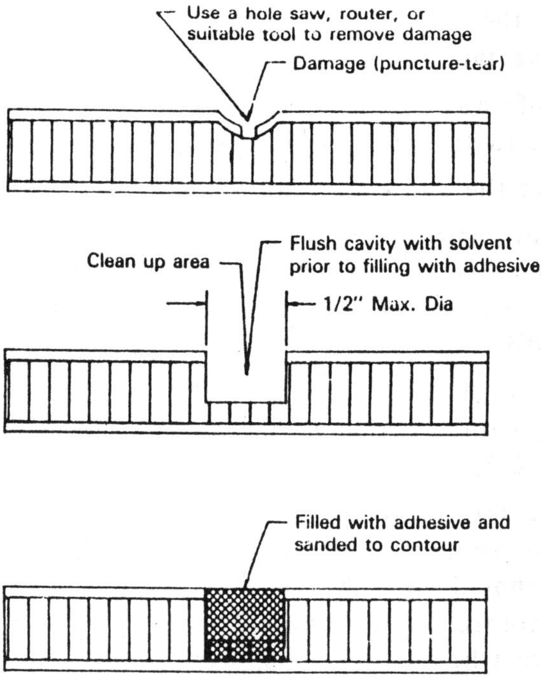

Figure 8-5: Repair of metal-faced structure containing dent and minor core damage.

4. Mix the aerodynamic smoother in accordance with the manufacturer's instructions.
5. Fill the hole over flush with the aerodynamic smoother cover with masking tape and cure in accordance with the manufacturer's instructions.
6. Refinish as necessary to obtain a smooth surface, prime and paint as required.

Method No. 8.6: Repair of Metal Faced Structures Containing Major Core Damage (Hole on One Skin Only)

The clean out should be at least two inches from the panel edge or a beveled edge.

Method of Repair:

1. Cover the damage with a solvent resistant pressure sensitive tape.
2. Mask off an area approximately four inches larger on all sides than the area to be repaired.
3. Remove the paint as described in chapter 4.
4. Lay out, using a scribe, the skin and core area to be removed. The layout area should have 1/8 inch minimum radius corners and the scribe lines should not extend past the radii.
5. Drill four 1/4 inch minimum holes in the corners of the layout, being careful not to drill to the bottom skin (See Figure 8-6).
6. Cut the skin between the drilled holes using an appropriate tool (router, skin saw or skin knife). If the damage is small enough the skin may be cut with a hole saw. A depth control should be used on all power tools to protect the core while cutting the skin.
7. Cut through the core with a sharp knife to the opposite skin following the cut in the skin.

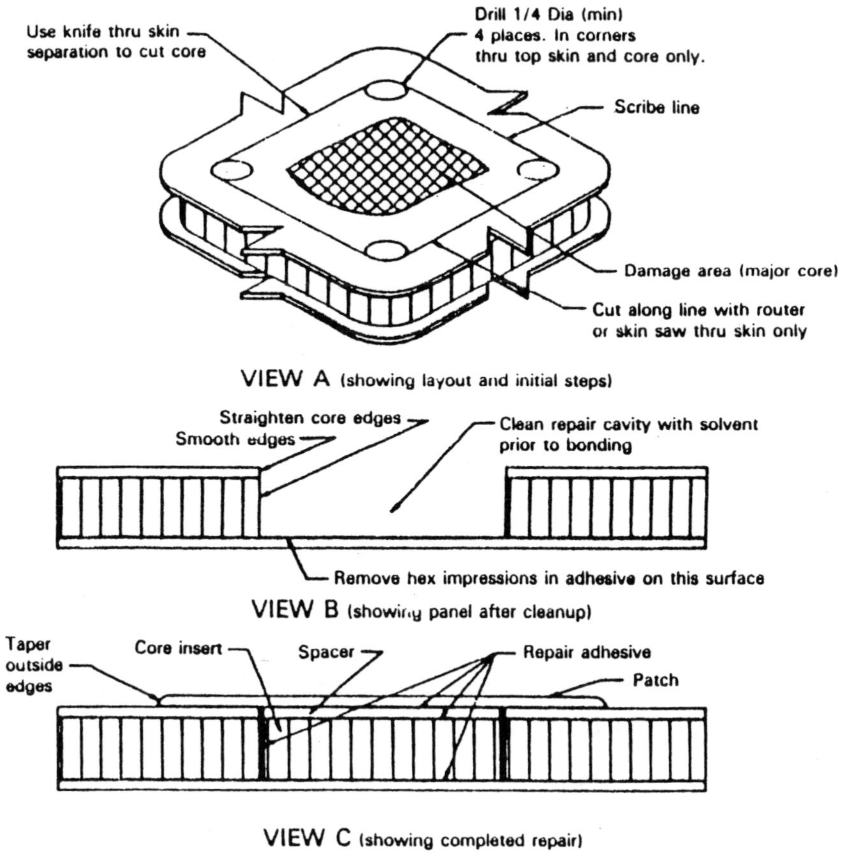

Figure 8-6: Repair of metal-faced structure containing major core damage (one skin only).

8. Carefully remove the damaged skin so as not to damage the skin outside of the removal area.

9. Remove and clean out the affected honeycomb area. Needle nose and duck bill pliers have been found useful for this purpose. After removal of the honeycomb core the hexagonal impressions should be removed from the adhesive adhered to the bottom skin. This may be done by a light sanding provided that the bottom skin is supported on a solid base. The adhesive need not be removed if it does not peel from the skin. A rotary file, wire brush or

sanding disc may be used if it is necessary to remove the adhesive. Care should be taken that excessive heat does not cause warping of the bottom skin. Excessive pressure without external support will cause delamination.

10. Smooth off sharp edges on the top skin with a file or suitable material.
11. Fabricate a replacement honeycomb insert from equal material to fit into the removed section. Keep the core ribbon direction the same as in the original structure. Straighten all core cells in the repair area and in the core insert to insure a maximum bond.
12. Clean the honeycomb insert and the cavity with trichloroethane and blow dry with clean, dry, oil free compressed air.
13. Mix a repair adhesive (See Table 2-1) in accordance with the manufacturer's instructions.
14. Apply the adhesive to the bottom and sides of the cavity and place the core insert into position. Fill all open spaces between the insert and the cavity with adhesive.
15. Apply a curing pressure and cure in accordance with the manufacturer's instructions.
16. Fabricate a spacer of the same material and thickness as the original skin which is large enough to fit into the damaged area (standard sheet metal flush skin patch insert gaps should be maintained). If core is available which will fill the cavity flush with the top of the skin, the spacer may be omitted. (See Figure 8-6, View B.)
17. Fabricate a patch from the same material and thickness as the skin which is large enough to overlap the damaged area by at least two inches on all sides. Bevel the outer edges and radius the corners. The two inch overlap may be reduced if rivets are to be used in conjunction with the bonding, provided that standard rivet practices and minimum edge distances are maintained (See Table 3-4).
18. Remove excessive cured adhesive from the previous

bonding operation, clean the core area and the surrounding area with trichloroethane spray or a dampened clean white cotton cloth.

19. Sand the surface of the panel spacer and the patch using No. 400 grit sandpaper. Use prebonded skin when available. Prebonded skin when available. (Prebonded skin are skins which have been chemically cleaned and to which a layer of structural film adhesive and a nylon peel ply have been previously bonded.) Remove the peel ply.
20. Mix the repair adhesive (See Table 2-1) in accordance with the manufacturer's instructions and apply the adhesive to the core insert surface and the spacer.
21. Place the spacer into the hole, apply adhesive to the top of the spacer, the patch and the skin surface.
22. Center the patch over the area of repair and apply pressure using a vacuum bag. Cure in accordance with the manufacturer's instructions.
23. Refinish as necessary, prime and paint as required.

Method No. 8.7: Repair of Metal Faced Structures Containing Major Core Damage (Hole in Both Skins)

The cleanout area should be at least two inches from the panel edge or a beveled edge.

Method of Repair:

1. Cover the damage with a solvent resistant pressure sensitive tape.
2. Mask off an area approximately four inches larger on all sides than the area to be repaired.
3. Remove the paint from the masked off area in accordance with the methods given in Chapter 4.
4. On the backside, using a scribe, layout the area to be removed. The layout area should have 1/8 inch minimum

radius corners and the scribe lines should not extend past the radii.

5. Drill four 1/4 inch minimum diameter holes in the corners of the layout being careful not to drill into the other skin. (Figure 8-7.)

6. Cut the skin between the drilled holes using a suitable tool such as a router, skin saw or skin knife. If the damage is small enough a hole saw may be used. A depth

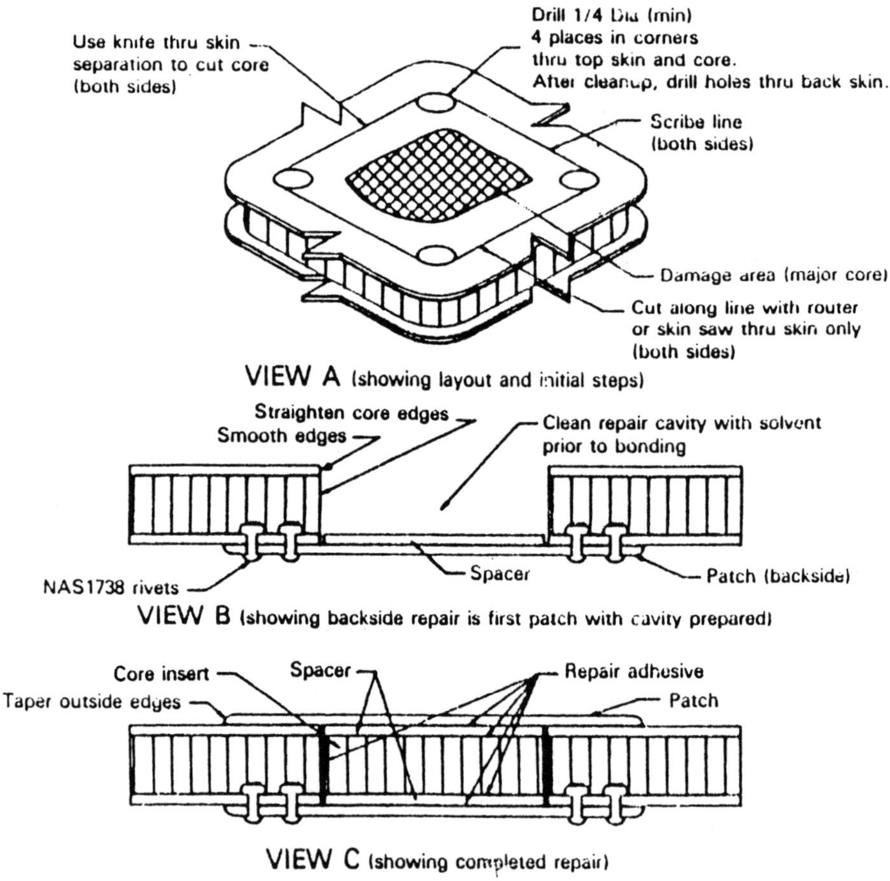

Figure 8-7: Repair of metal-faced structure containing major core damage (Hole Both Skins).

control should be used on all power tools to protect the core.

7. Cut through the core with a sharp knife to the opposite skin, following the cut in the skin.

8. Carefully remove the damaged skin from the honeycomb so as not to damage the skin outside of the removal area.

9. Remove and clean out the affected core. Needle nose and duck bill pliers have been found useful for the removal of damaged core.

10. Smooth all sharp edges in the bottom skin with a file or suitable grinder.

11. Support the opposite skin from the outside and drill four 1/4 inch minimum diameter holes (same size as in Step 5.).

12. Fabricate a spacer of the same material and thickness as the skin which is large enough to fit into the damaged area (standard sheet metal flush patch insert gaps should be maintained). Also fabricate a patch from the same material that is large enough to overlap the spacer by two inches on all sides. Sand the spacer on both sides with 400 grit sandpaper. Bevel the outer edges of the patch and radius the corners. Sand the bottom side of the patch with 400 grit sandpaper.

13. Bond the spacer to the center of the patch with the repair adhesive (See Table 2-1). Cure in accordance withthe manufacturer's instructions.

14. Bond and rivet the patch to the backside of the structure with the insert in the cut out area. (See Figure 8-7, View B.)

15. Fabricate a replacement honeycomb core insert from equal material to fit the removed section. Keep the ribbon direction the same as in the original structure. Straighten all edges of the core cells in the insert and the repair area.

16. Clean the honeycomb core insert and the cavity with tri-

chloroethane and dry with clean, dry oil free compressed air.

17. Mix and apply a layer of repair adhesive (See Table 2-1) to all sides of the repair cavity and to the edges of the core insert. Place the insert in position and fill all voids in the bondline between the core insert and the structure with adhesive. Apply curing pressure and cure in accordance with the adhesive manufacturer's instructions.

18. Fabricate a spacer and a patch of the same materials and thickness as the original skin. The insert shall be large enough to fit into the damaged area. (Standard sheet metal flush patch insert gap shall be maintained). If core is available which will fill the cavity flush with the top of the skin, the spacer may be omitted. The patch should be large enough to cover the insert by at least two inches on all sides. The two inch minimum overlap may be reduced if rivets are used in conjunction with the adhesive and if standard rivet practice and minimum edge distances are maintained (See Table 3-4). Bevel the outer edges and radius the corners of the patch.

19. Remove the excess adhesive from the previous bonding operation and clean the core area with a trichloroethane spray. Sand both surfaces of the spacer and the bottom surface of the patch with No. 400 grit sandpaper. Use prebonded skins when available. Prebonded skins are skins which have been chemically cleaned and to which a layer of a structural adhesive and a peel ply have been previously bonded. Remove the peel ply before applying the repair adhesive.

20. Mix repair adhesive (See Table 2-1) in accordance with the manufacturer's instructions and apply to the core insert surface and to the spacer. Place the spacer into the hole.

21. Apply adhesive to the top surface of the spacer and to the patch and skin surface. Center the patch over the area of repair and secure it with masking tape.

22. Apply curing pressure and cure in accordance with the adhesive manufacturer's instructions. NOTE: If one skin is nonmetallic, use the appropriate method for a composite skin repair.
23. Refinish as necessary, prime and paint as required.

Method No. 8.8: Repair of Metal Faced Structures Containing Edge Voids

Edge voids should not be considered negligible damage and any panel that has lost its edge seal should be resealed.

Method of Repair:

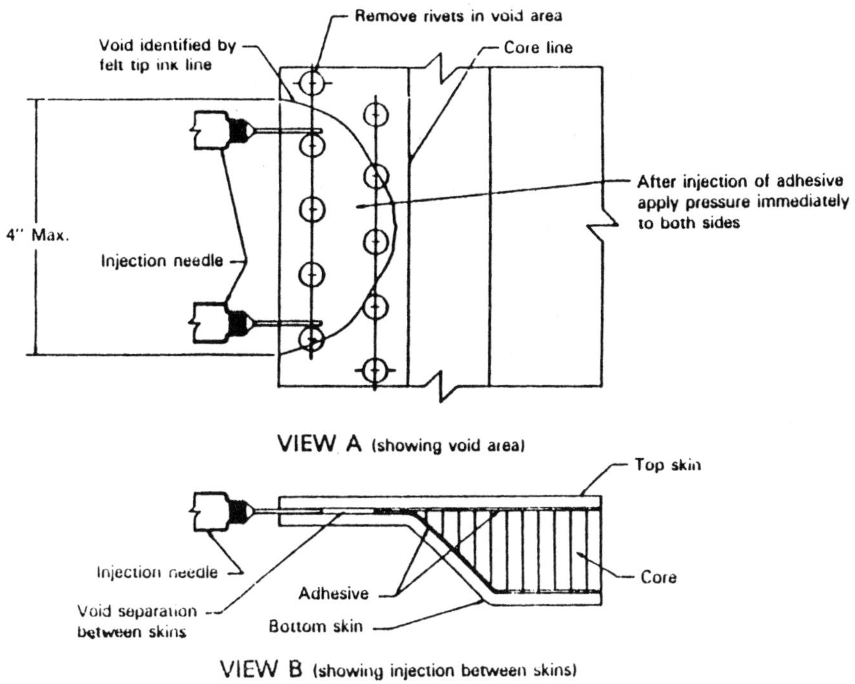

Figure 8-8: Repair of metal-faced structure containing edge voids.

1. Using a drill bit the same size as the rivet shank, drill through the center of the rivet head to the panel surface. Using a drift punch, pop off the rivet head then lightly drive out the rivet shank using a drift punch. NOTE: For blind lock bulb rivets, a rivet shaver or a grinder may be used in lieu of drilling and popping off of the head.
2. Mix a repair adhesive (See Table 2-1) in accordance with the manufacturer's instructions.
3. Using an injection needle inject the adhesive directly into the void from a separated edge. (See Figure 8-8.) Immediately apply pressure to the void side of the panel to allow the excess adhesive to flow from the holes.
4. Cure the adhesive under pressure, in accordance with the adhesive manufacturer's instructions.
5. Seal the edges with sealing compound (See Table 2-1).

9

LARGE AREA DEPOT REPAIR OF METAL-FACED STRUCTURES

Large area repairs cannot generally be made at field location. This is, to a large extent, controlled by the availability of the proper equipment such as autoclaves, bonding tools, alignment fixtures and the cleaning facilities required. These type repairs must be made at a depot or factory having the required equipment.

Method No. 9.1: Large Area Major Damage (Hole in One Skin Only).

This method covers the use of structural film adhesives and core splice adhesives in the repair of large area major damage.

Method of Repair:

The cleanout area should be at least two inches from the panel edge or a beveled edge.

1. Cover the damage with a solvent resistant pressure sensitive tape.

2. Mask off an area approximately four inches larger on all sides than the area to be repaired.
3. Remove the paint using an appropriate method from Chapter 4.
4. Lay out using a scribe, the skin and core area to be removed. The layout area should have 1/8 inch minimum radius corners and the scribe lines should not extend past the radii.
5. Drill four 1/4 inch minimum holes in the corners of the layout, being careful not to drill to the bottom skin. (Figure 8-6.)
6. Cut the skin between the drilled holes using an appropriate tool (router, skin saw or skin knife). If the damage is small enough the skin may be cut with a hole saw. A depth control should be used on all power tools to protect the core while cutting the skin.
7. Cut through the core with a sharp knife to the opposite skin following the cut in the skin.
8. Carefully remove the damaged skin so as not to damage the skin outside of the removal area.
9. Remove and clean out the affected honeycomb area. Needle nose and duck bill pliers have been found useful for this purpose. After removal of the honeycomb core, the hexagonal impressions should be removed from the adhesive adhered to the bottom skin. This may be done by lightly sanding provided that the bottom skin is supported on a solid base. The adhesive need not be removed if it does not peel from the skin. A rotary file, wire brush or sanding disc may be used if it is necessary to remove the adhesive. Care should be taken that excessive heat does not cause warping of the bottom skin. Excessive pressure without external support will cause delamination.
10. Smooth off sharp edges on the top skin with a file or suitable grinder.
11. Fabricate a replacement honeycomb insert from equal

material to fit into the removed section. Keep the core ribbon direction the same as in the original structure. Straighten all core cells in the repair area and in the core insert to insure a maximum bond.

12. Clean the honeycomb insert and the cavity with trichloroethane and blow dry with clean, dry, oil free compressed air.

13. Cut a section of the structural film adhesive (See Table 2-1) to fit the bottom of the cavity.

    NOTE: Structural film adhesives must be stored at temperatures below zero degrees F. They should be removed from storage at least 12 hours before they are needed to allow the adhesive to come to room temperature before removal of their protective packaging.

14. Remove the separator film from one side of the adhesive film and apply the adhesive to the bottom of the cavity. Remove the remaining separator film.

15. Cut a strip of core splice adhesive (See Table 2-1) to fit the sides of the core insert. Remove the separator film from one side and apply the adhesive by pressing it to the edge of the core. Remove the other separator film. Place the insert into the cavity. Apply curing pressure and cure in accordance with the adhesive manufacturer's instructions.

16. Fabricate a spacer of the same material and thickness as the original skin which is large enough to fit into the damaged area (standard sheet metal flush skin patch insert gaps should be maintained). If core is available which will fill the cavity flush with the top of the skin, the spacer may be omitted. (See Figure 8-7, View C.)

17. Fabricate a patch from the same material and thickness as the skin which is large enough to overlap the damage area by at least two inches on all sides. Bevel the outer edges and radius the corners. The two inch overlap may be reduced if rivets are to be used in conjunction with the bonding provided that standard rivet practices and

minimum edge distances are maintained (See Table 3-4).

18. Remove excessive cured adhesive from the previous bonding operation, clean the core area and the surrounding area with trichloroethane spray or a dampened clean white cotton cloth.

19. Sand the surface of the panel spacer and the pacer using No. 400 grit sandpaper. Use prebonded skin when available. Prebonded skin are skins which have been chemically cleaned and to which a layer of structural film adhesive and a nylon peel ply have been previously bonded. Remove the peel ply.

20. Cut two pieces of film adhesive, one the size of the insert and the other the size of the patch. Remove the separator sheets and apply the adhesive to the insert and the patch.

21. Place the insert into the hole, adhesive side down.

22. Center the patch over the area of repair and apply pressure using a vacuum bag. Cure in accordance with the manufacturer's instructions.

23. Refinish as necessary, prime and paint as required. Seal edges with sealing compound (See Table 2-1).

Method No. 9.2: Large Area Major Damage in Both Skins

Method of Repair:

1. Cover the damage with a solvent resistant pressure sensitive tape.

2. Mask off an area approximately four inches larger on all sides than the area to be repaired.

3. Remove the paint from the masked off area using a method defined in Chapter 4.

4. On the backside, using a scribe, layout the area to be removed. The layout area should have 1/8 inch minimum radius corners and the scribe lines should not extend past the radii.

5. Drill four 1/4 inch minimum diameter holes in the corners of the layout being careful not to drill into the other skin.
6. Cut the skin between the drilled holes using a suitable tool such as a router, skin saw or skin knife. If the damage is small enough a hole saw may be used. A depth control should be used on all power tools to protect the core.
7. Cut through the core with a sharp knife to the opposite skin, following the cut in the skin.
8. Carefully remove the damaged skin from the honeycomb so as not to damage the skin outside of the removal area.
9. Remove and clean out the affected core. Needle nose and duck bill pliers have been found useful for the removal of damaged core.
10. Smooth all sharp edges in the bottom skin with a file or suitable grinder.
11. Support the opposite skin from the outside and drill four 1/4 inch minimum diameter holes (same size as in Step 5.).
12. Fabricate a spacer of the same material and thickness as the skin which is large enough to fit into the damaged area (standard sheet metal flush patch insert gaps should be maintained). Also fabricate a patch from the same material that is large enough to overlap the spacer by two inches on all sides. Sand the spacer on both sides with 400 grit sandpaper. Bevel the outer edges of the patch and radius the corners, sand the bottom side of the patch with 400 grit sandpaper.
13. Cut a piece of film adhesive the size of the patch, remove one separator film and place the adhesive on the sanded side of the patch. Remove the other separator film and center the spacer on the patch.
14. Bond the patch to the backside of the structure with the insert in the cut out area. Rivets may be used to hold patch in place. Cure in accordance with the adhesive manufacturer's instructions.

15. Fabricate a replacement honeycomb core insert from equal material to fit the removed section. Keep the ribbon direction the same as in the original structure. Straighten all edges of the core cells in the insert and the repair area.
16. Clean the honeycomb core insert and the cavity with trichloro- ethane and dry with clean, dry oil free compressed air.
17. Cut a section of structural film adhesive to fit the bottom of the cavity. Remove the separator sheet from one side of the adhesive and apply the adhesive to the bottom of the cavity, remove the other separator sheet.
18. Cut a strip of core splice adhesive (See Table 2-1) to fit the sides of the core insert. Remove the separator film from one side of the core splice adhesive and apply the adhesive by pressing it to the edges of the core. Remove the other separator sheet and place the insert into the cavity. Apply curing pressure and cure in accordance with the manufacturer's instructions.
19. Fabricate a spacer and a patch of the same materials and thickness as the original skin. The insert must be large enough to fit into the damaged area. (Standard sheet metal flush patch insert gap shall be maintained.) If core is available which will fill the cavity flush with the top of the skin, the spacer may be omitted. The patch should be large enough to cover the insert by at least two inches on all sides. The two inch minimum overlap may be reduced if rivets are used in conjunction with the adhesive, provided standard rivet practice and minimum edge distances are maintained. (See Table 3-4.) Bevel the outer edges and radius the corners of the patch.
20. Remove excess adhesive from the previous bonding operation and clean the core area with a trichloroethane spray. Sand both surfaces of the spacer and the bottom surface of the patch with No. 400 grit sandpaper. Use prebonded skins when available. Prebonded skins are skins

which have been chemically cleaned and to which a layer of a structural adhesive and a peel ply have been previously bonded. Remove the peel ply before applying the repair adhesive.

21. Cut two pieces of the film adhesive, one the size of the insert and the other the size of the patch. Remove the separator film and apply the adhesive to the spacer and the patch. Place the insert into the hole adhesive side down.
22. Center the patch over the repair area.
23. Apply curing pressure and cure in accordance with the adhesive manufacturer's instructions.

    NOTE: If one skin is nonmetallic use the appropriate method for a composite skin repair.
24. Refinish as necessary, prime and paint as required.

REBUILD OR FABRICATION OF A METAL FACED STRUCTURE.

When a metal faced adhesive bonded structure can no longer be repaired because of the extent of the damage it may be possible to rebuild the structure using some of the original components. Rebuilding involves the removal of the damaged material, the fabrication of new parts and the rebuilding of the structure using chemical treatments for the preparation of the metal parts and the use of heat curing structural adhesives.

Method No. 9.3: Rebuild of a Metal-Faced Structure

This method is applicable when the damage exceeds the limits of depot repair and cover the replacement of up to the entire skin and up to the entire core if necessary.

Method of Repair:

1. Remove the delaminated or damaged skin taking care not to damage any undamaged core. Various types of mechani-

cal equipment are available to cut, slit, rout or saw through the metallic skin. Equipment commonly used includes the high speed router, skin saw or slotting saw. A sharpened putty knife may be used to lift the edge of the skin, after which it may be gripped with a pair of pliers, vise grips or a slotted Tee bar tool. When attempting to remove a skin, be careful to pull slowly and be aware that there may be a sudden release of the bond. The coiled skin may attempt to uncoil and the edges will be sharp. If the adhesive has a high peel strength dry ice (solid carbon dioxide) may be applied to the skin surface to embrittle the adhesive and lower its peel strength. An effective procedure is to cover the skin with dry ice, then cover the entire part with an insulation blanket. After about 30 minutes the blanket and dry ice are removed and the skin is peeled off.

NOTE: Dry ice should be handled with insulated gloves and safety glasses should be worn. Serious frostbite or burns may result if dry ice comes in contact with the skin or eyes. Provide ventilation for the carbon dioxide in accordance with local safety regulations. In case of dry ice burns obtain medical assistance immediately.

2. If after the skin has been removed oil or fuel is detected, the assembly should be immersed in trichloroethane vapors and thoroughly flushed. The assembly shall then be dried with clean, dry, oil free compressed air. Care should be taken during drying with compressed air to avoid damaging the honeycomb core. If moisture is detected in the panel instead of fuel, it is removed by the application of heat from heat lamps or other heat source. The temperature should not exceed 180 degrees F.

3. Remove the damaged honeycomb core using a suitable tool which will not damage the undamaged core. Make vertical cuts through the damaged core using a sharp knife, sharpened putty knife, or a router. Remove the core down to the adhesive bond line. If the adhesive bond to the lower skin is intact the adhesive should not be removed.

Any hexagonal imprints or surface gloss on the adhesive should be removed with a disc sander. Care should be taken so as not to penetrate the adhesive layer and expose the base skin. An adhesive layer that is well adhered to the metal skin will provide an excellent surface for the new bond.

4. Remove core chips and debris with a vacuum cleaner or other suitable means. Clean the exposed surfaces with trichloroethane to remove traces of sanding dust.

5. Cut and prefit replacement core of the same material and density as the removed core. The core should be fabricated so that the ribbon direction matches that of the original core. Special equipment and tools may be required when contoured honeycomb shapes are involved, see MIL-HDBK-337. Thicknesses must be maintained within 0.010 inch. Burrs on honeycomb core are unacceptable. Splices on core should be butt joints using a core splice adhesive (See Table 2-1.). The core must be cleaned by vapor degreasing and dried using hot air.

6. Fabricate and prefit any internal doublers that may be required. Chemically clean using an appropriate process.

7. Fabricate and prefit a new skin of the same material and thickness as the original. Prepare the skin using an appropriate cleaning process.

8. Prime the metal surface as required. The primer shall be applied to the faying surface and cured in accordance with the manufacturer's instructions. Maintain primer thickness within the tolerances established by the primer manufacturer unless other tolerances have been established for the particular part. If the primer has been stored in a refrigerator it must be allowed to warm to room temperature before the container is opened to prevent condensation of moisture. The primer shall be agitated prior to and during use to keep the solid content in suspension. Primed and cured parts must be protected from contamination and excessive handling.

9. Remove the adhesives to be used from cold storage and allow to equilibrate to room temperature.
10. Cut the film adhesive to fit the detail parts to be assembled. Cut the core splice adhesive to fit any splice joints that will have to be made in the core.
11. Remove the separator film from one side adhesive and apply the adhesive to the faying surfaces of the parts to be assembled, being careful to prevent air entrapment. For large parts lay the adhesive from the center of the part and work towards the edges. Leave the separator sheet on the side of the adhesive exposed to the atmosphere as a temporary protective cover. If the adhesive film is used in cut sections there shall be no gaps between pieces of adhesive. Overlapping of the adhesive film is not recommended unless required for tolerance build-ups in beveled sections or when rigidized skins are used.
12. Install the honeycomb core replacement into the panel assembly. Apply core splice adhesive between butt joints in the core. Core splice adhesive should be used between the core and edge members if applicable.
13. Install doublers as needed after removal of the separator sheet from the adhesive. Care should be taken in the assembling of the parts in the curing fixture or jig to assure that the adhesive is not disturbed.
14. Remove the separator sheet from the adhesive on the skin and carefully place the skin on to the panel.
15. Place the part in a vacuum bag and cure in accordance with the adhesive manufacturer's instructions.
16. Refinish, prime and paint as necessary.

FABRICATION OF NEW STRUCTURES

When a structure can not be repaired, rebuilt, or replaced from stock a new structure may be fabricated. This is done in accordance with the original drawings and specifications and their latest revisions. The materials of fabrication and the processes used shall be the same as in the original structure.

10

REPAIR OF SOLID CORE STRUCTURES

This chapter provides instructions for making repairs on adhesive bonded structures which contain solid core materials such as foams and low density woods such as balsa wood. These repairs are essentially patch type repairs and the limits to their use is the responsibility of the cognizant engineering agency for the structure being repaired.

The solid core structures are sandwich type construction using thin skin materials bonded to a core of either a foam material or end grain balsa wood. This type of structure is often used to obtain a stiff, light weight panel that has excellent thermal and electrical insulating characteristics, and outstanding compressive strengths. Strength and rigidity, light weight, and smooth surfaces are among the desirable features of this type of structure.

The foam materials most frequently used are closed cell polyurethane foams and polystyrene foams. Other foam materials such as low density syntactic film cores are taking their place in the structural materials. The methods outlined in this chapter have been developed for the urethane and styrene foams.

Balsa wood cores are generally used for their light weight and the high compressive strengths exhibited when balsa wood is loaded parallel to the grain.

As in the case of honeycomb core sandwich construction, the solid core sandwich construction is analogous to the concept of the I-beam where the skins or face sheets act as the flange and the core acts as the web.

FOAM CORE SANDWICH STRUCTURES

Method No. 10.1: Repair of a Puncture - Flush Patch Method (Ref. 10-1)

Method of Repair:

1. Mark a rectangular area that is one (1) inch larger on all sides than the size of the puncture and its surrounding dent or other damage.
2. From this outline determine the size of the rectangular patch panel and cut it to size using the same material and thickness as the original panel material. (See Figure 10-1.)
3. Place the patch panel over the repair area, centrally located within the original outline, and scribe its outline on the panel to be repaired.
4. Using a Dotco saw, pin router or other suitable tool, cut through the damaged skin along the inside edges of the scribed outline.
5. Using a putty knife, sharp knife or other suitable tool, cut through the damaged foam following the previously made cuts in the skin.
6. Remove the punctured skin and foam material.

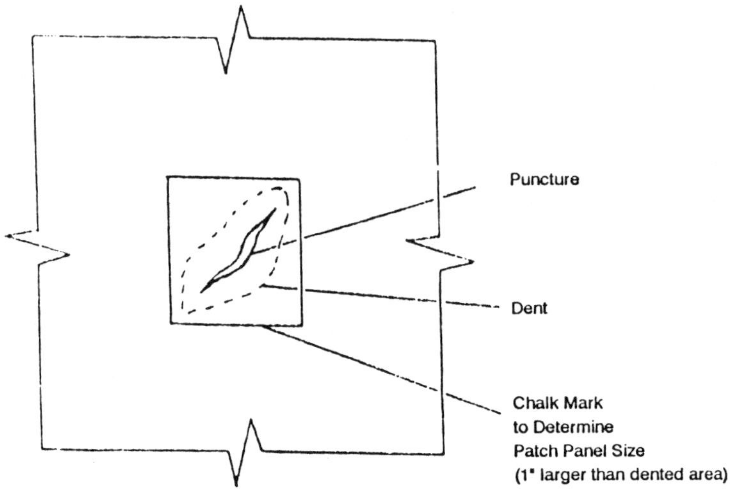

Figure 10-1: Layout of damaged area of a foam core panel.

7. Using a hand file, trim the edges of the cutout so that they are straight, and so that the cutout is 1/16" larger than the patch panel. (See Figure 10-2.)
8. Using a suitable off-set tool such as a saw blade or other under-cutting router bit, remove the foam and adhesive from the back side of the edges of the cutout. (See Figure 10-2, Section A-A.)
9. Prepare a replacement foam insert that is the same size (length and width as the cut out. The fit should be snug. The replacement should be of the same thickness and material as the original.
10. Prepare 4 framing strips of the same material and thickness as the ruptured skin. These strips shall be 1" wide and the lengths shall be as follows: 2 ea. 1" longer than the patch panel width, 2 ea. 1" shorter than the patch panel height.
11. Compress or remove foam on the four edges of the replacement insert so that the surface of the framing strips is even with the face of the foam. (See Figure 10-3, Section B-B.)

12. Remove any paint from the cutout edges using a method given in Chapter 4.
13. Layout and drill #30 holes at spacing shown in Figure 10-4.
14. Layout and drill #30 holes in the patch at the spacings shown in Figure 10-5.
15. Prefit the foam insert into the cutout area, check for snug fit, remove and set aside.
16. Place one of the longer framing strips at the top edge of the cutout as shown in Figure 10-5.
17. Transfer drill through the framing strip, using the holes along the cutout edge as a guide, and secure in place using Cleco clamps. Continue this step using the two shorter strips on the sides and the other long strip at the bottom. Make sure that all strips are tightly butted against each other.
18. Center the patch panel in the cutout and transfer drill through the framing strips using the patch panel holes as a guide, and secure with Cleco clamps where necessary.
19. Countersink all holes so that rivet heads are flush with, or slightly below, the skin surface. (Slightly below flush is preferred.
20. Remove patch panel and framing strips and deburr all holes. Place each part so that it can be returned to its original location.
21. Remove all foreign materials such as dust, chips, savings, etc., from the cutout area and the surrounding panel surface.
22. Using safety solvent, thoroughly clean all surfaces of the patch panel, framing strips and the surrounding panel surface.
23. Mix a repair adhesive (See Table 2-1) in accordance with the manufacturer's instructions.
24. Apply adhesive to all inside surfaces of the cutout area and install the replacement foam core.

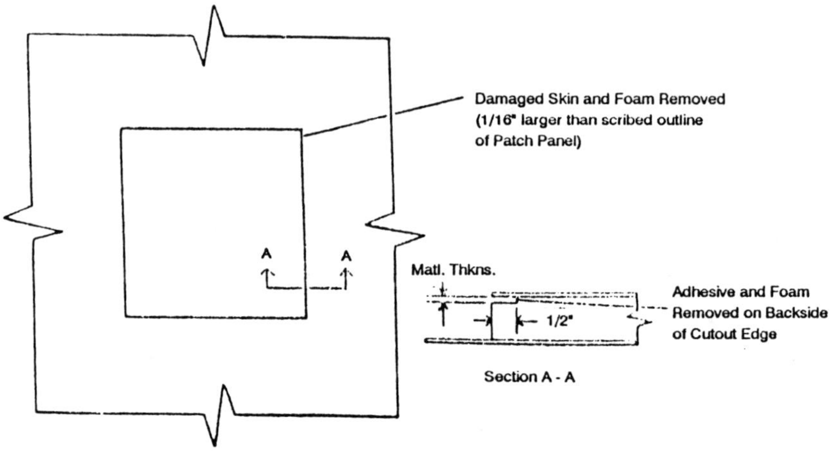

Figure 10-2: Cutout of damaged area in a foam panel.

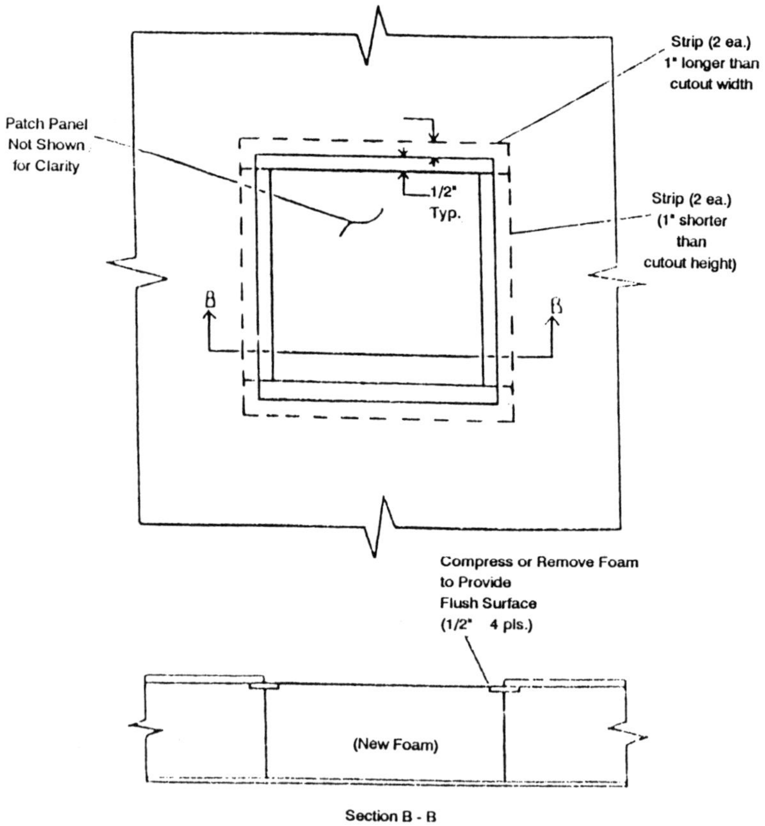

Figure 10-3: Framing strips in repair of foam core panels.

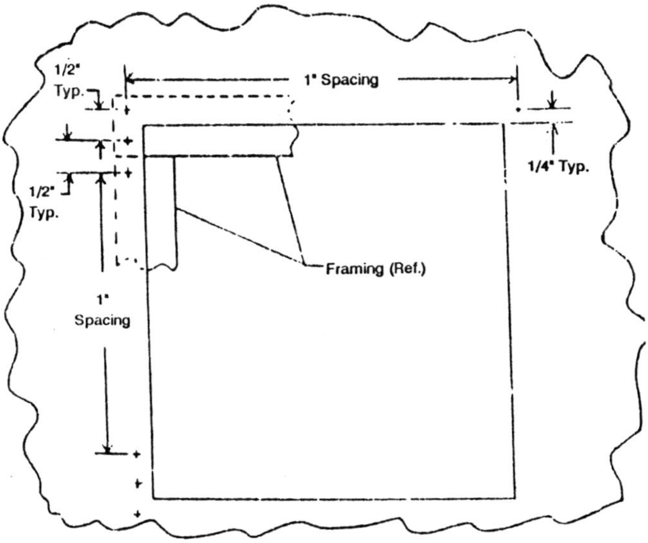

Figure 10-4: Hole location in panel surface.

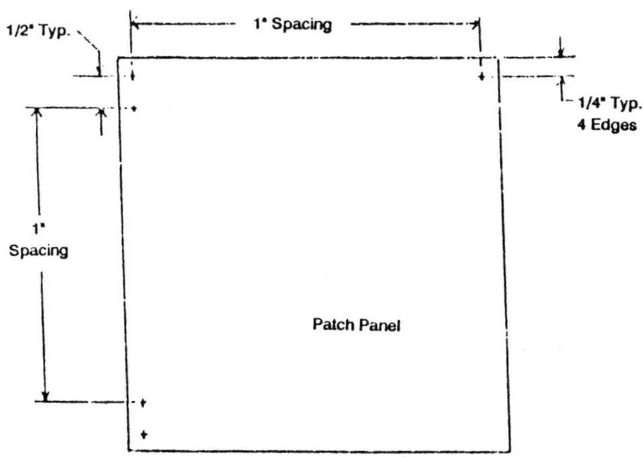

Figure 10-5: Hole spacing in patch panel.

25. Lightly coat the framing strips with adhesive and slide into place around the edges of the cutout.
26. Secure framing strips in place using countersunk rivets dipped in sealer (See Table 2-1), mixed in accordance with the manufacturer's instructions.
27. Apply adhesive to the remaining surface of the replacement foam core and the back side of the patch panel.
28. Install the patch panel using countersunk rivets dipped in sealer.
29. Remove all excess adhesive squeezed out during the installation process and cure the adhesive in accordance with the manufacturer's instructions.
30. Fill the 1/32" gap between the edges of the patch panel and the cutout. Also cover the heads of all countersunk rivets.
31. Allow the adhesive to cure and the sand smooth using an oscillating sander.
32. Refinish as required.

Method No. 10.2: Repair of a Through Hole in a Foam Core Panel - Flush Patch Method (Ref. 10-1)

Method of Repair:

1. Mark a rectangular area that is one (1) inch larger on all sides than the size of the puncture and its surrounding dent or other damage.
2. From this outline determine the size of the rectangular patch panel and cut it to size using the same material and thickness as the original panel material. (See Figure 10-6.)
3. Place the patch panel over the repair area, centrally located within the outline, and scribe its outline on the panel to be repaired.
4. Using a router or other suitable tool, cut through the

damaged skins along the inside edges of the scribed outline.

5. Remove the punctured skin and foam material.
6. Using a hand file, trim the edges of the cutout so that they are straight, and so that the cutout is 1/16" larger than the patch panel. (See Figure 10-.7)
7. Using a suitable off-set tool such as a saw blade or other-under cutting router bit, remove the foam and adhesive from the back side of the edges of the cutout. (See Figure 10-7, Section A-A.)
8. Prepare a replacement foam insert that is the same size (length and width) as the cutout. The fit should be snug. The replacement should be of the same thickness and material as the original.
9. Prepare 4 framing strips of the same material and thickness as the ruptured skin. These strips shall be 1" wide and the lengths shall be as follows: 2 ea. 1" longer than the patch panel width, 2 ea. 1" shorter than the patch panel height.
10. Compress or remove foam on the four edges of the replacement insert so that the surface of the framing strips is even with the face of the foam. (See Figure 10-8, Section B-B.)
11. Remove any paint from the edge of cutout edges using a method described in Chapter 4.
12. Layout and drill #30 holes at spacing shown in Figure 10-9 on both sides of the cutout.
13. Layout and drill #30 holes in the patch at the spacings shown in Figure 10-10.
14. Prefit the foam insert into the cutout area, check for snug fit, remove and set aside.
15. Place one of the longer framing strips at the top edge of the cutout as shown in Figure 10-8.
16. Transfer drill through the framing strip, using the holes

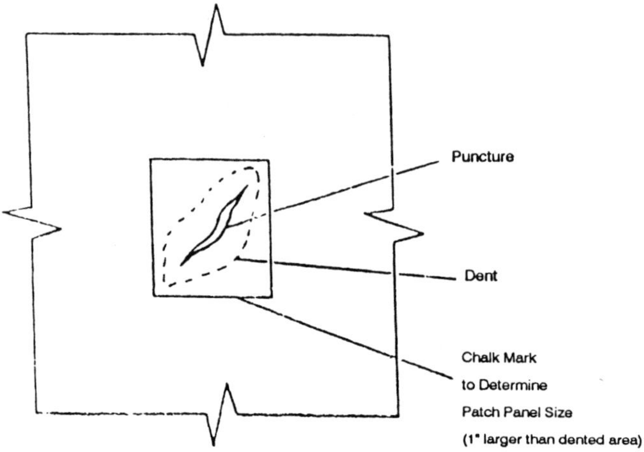

Figure 10-6: Lay-out of damaged area.

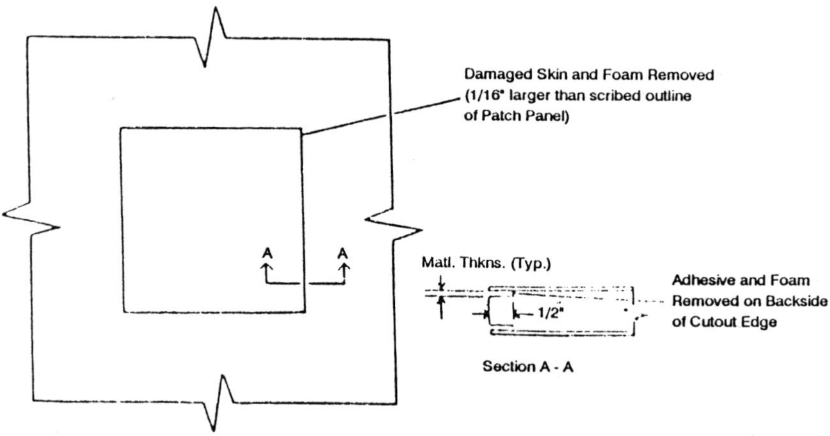

Figure 10-7: Cutout of damage area - through hole foam panel.

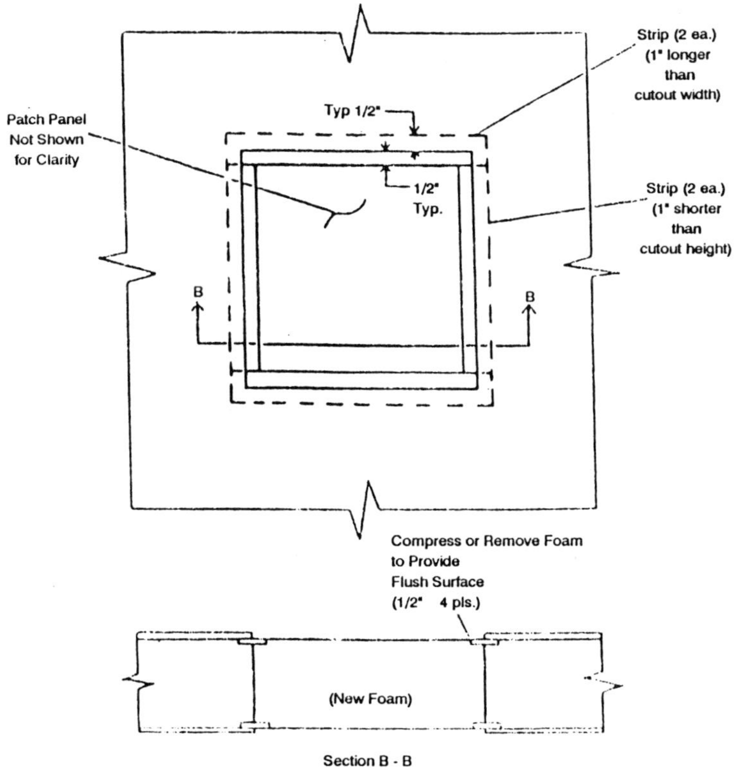

Figure 10-8: Framing strips in repair of through hole foam panel.

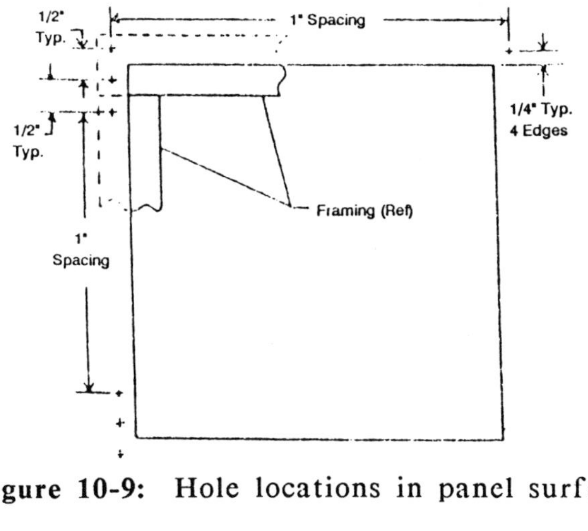

Figure 10-9: Hole locations in panel surface.

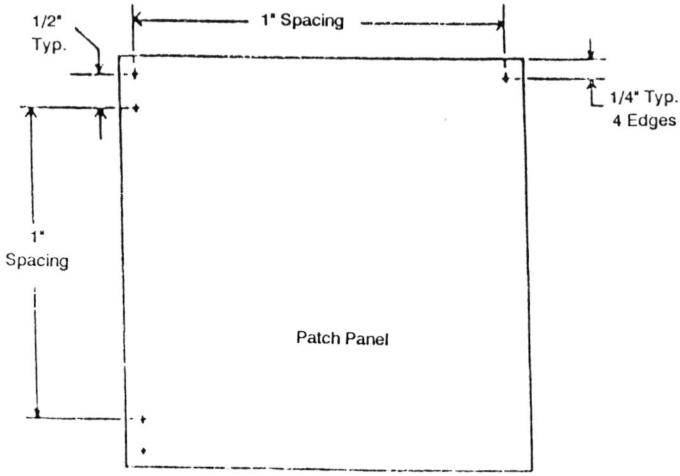

Figure 10-10: Hole spacing in patch panel.

along the cutout edge as a guide, and secure in place using Cleco clamps. Continue this step using the two shorter strips on the sides and the other long strip at the bottom. Make sure that all strips are tightly butted against each other.

17. Center the patch panel in the cutout and transfer drill through the framing strips using the patch panel holes as a guide, and secure with Cleco clamps where necessary.
18. Countersink all holes so that rivet heads are flush with, or slightly below, the skin surface. (Slightly below flush is preferred.
19. Remove patch panel and framing strips and deburr all holes. Mark each part so that it can be returned to its original location.
20. Repeat steps 15 through 19 for the other side of the cutout.

21. Remove all foreign materials such as dust, chips, savings, etc. from the cutout area and the surrounding panel surface.
22. Using safety solvent, thoroughly clean all surfaces of the patch panel, framing strips and the surrounding panel surface.
23. Mix a repair adhesive (See Table 2-1) in accordance with the manufacturer's instructions.
24. Lightly coat the external surface of the framing strips for one side of the cutout with adhesive and place around the edges of the cutout.
25. Secure framing strips in place using countersunk rivets dipped in sealer (See Table 2-1) mixed in accordance with the manufacturer's instructions.
26. Install the appropriate patch panel using countersunk rivets dipped in sealer.
27. Remove all excess adhesive squeezed out between the framing strips and the patch panel during the installation process.
28. Apply adhesive to all inside surfaces of the cutout area and install the replacement foam.
29. Lightly coat the remaining framing strips with adhesive and slide into place around the edges of the cutout.
30. Secure the framing strips in place using countersunk rivets dipped in sealer.
31. Apply adhesive to the remaining surface of the replacement foam core and the back side of the patch panel.
32. Install the patch panel using countersunk rivets dipped in sealer.
33. Remove all excess adhesive squeezed out during the installation process and cure the adhesive in accordance with the manufacturer's instructions.
34. Fill the 1/32" gap between the edges of the patch panel

and the cutout. Also cover the heads of all countersunk rivets.

35. Allow the adhesive to cure, then sand smooth using an oscillating sander.
36. Refinish as required.

Method No. 10.3: Repair of Minor Damage - External Patch Method (Ref. 10-2)

Method of Repair:

1. Bend edges of the puncture in below the surface of the unbroken face sheet. Do not allow broken edges to contact the opposite face.
2. Remove loose fragments of foam and dust.
3. Prepare a patch, using the same material as the original skin, large enough to extend at least one inch beyond the damaged area in all directions.
4. Center the patch over the damaged area, then lightly trace the outline of the patch on the panel surface.
5. Remove all the paint from panel surface within the outline by using a method given in Chapter 4.
6. Lay out rivet hole pattern around the edge of the patch. The rivet holes should be 1/4 to 1/2 inch in from the edge of the patch and one inch apart. Drill holes using a #11 or #12 drill bit. Use eye protection while drilling.
7. Center patch over the damaged area of the panel.
8. Using the holes in the patch as a guide, drill holes in the diagonally opposite corners. Then drill the other two corner holes. Cleco clamps may be used to hold the patch in place, or rivets may be temporarily inserted into each hole as it is drill to keep the patch from moving. Proceed to drill the remaining holes. Use a drill and pin as a stop by placing the drill inside the pin and setting the drill in

the chuck so that 7/16 to 9/16 inch of the drill is exposed when the pin is against the chuck jaws.

9. Remove the patch and remove all burrs from the drilled holes.

10. If the puncture is surrounded by extended cracks, stop drill the ends of the crack using a #30 drill.

11. Prepare the bonding surfaces of the panel and the patch for bonding by sanding until the oxide film is removed and a bright surface is exposed.

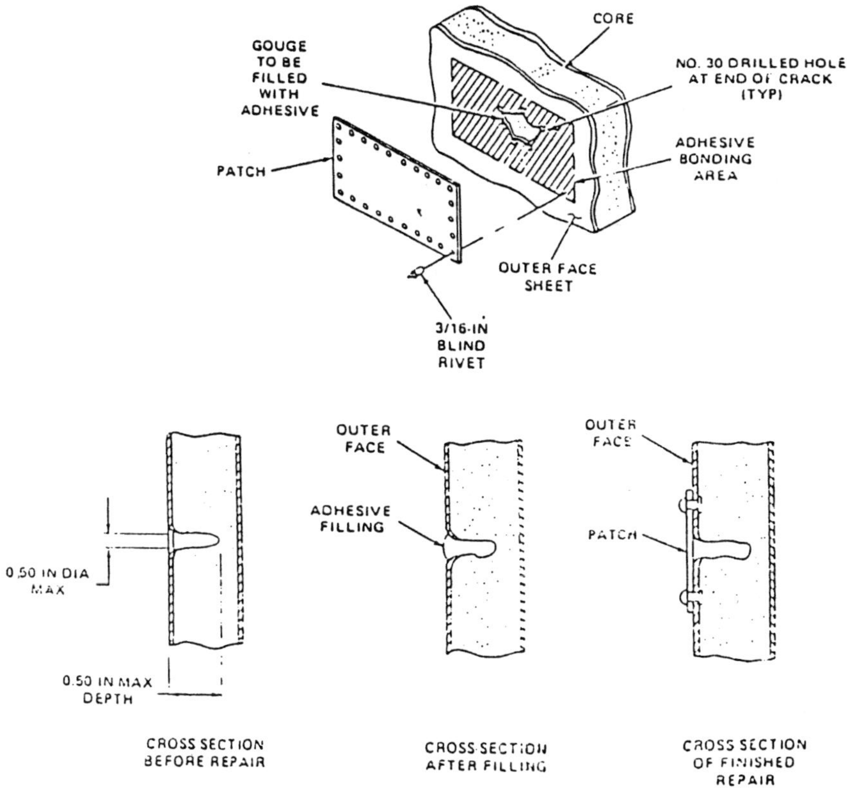

Figure 10-11: Repair of minor damage using the external patch method.

12. Inject sealant (Table 2-1) into the puncture (See Fig 10-11) and allow to harden.
13. After sealant has hardened, sand until sealant flush is flush with the surface of the panel.
14. Wipe panel surface and the patch surface with a clean white cotton cloth and safety solvent.
15. Apply a layer of repair adhesive (See Table 2-1) to the faying surface of both the panel and the patch. Position the patch in place with rivets which have been coated with the sealant.
16. Rivet the patch to the panel surface. When repairing an external skin, before installing rivets, inject sealant into drilled holes to prevent moisture from entering the panel. (See Figure 10-15 for a description of closed- end rivets used in these repairs.)
17. Wipe any excess adhesive from the surface of the repair area. Do not use solvents to remove adhesive.
18. Allow adhesive to cure in accordance with the manufacturer's instructions.
19. Coat the edges of the patch and the rivet heads with sealant.
20. Refinish as required.

Method No. 10-4: Repair of Puncture with Major Core Damage - External Patch Method (Ref. 10-2)

Method of Repair:

When damage involves one or both skins, extends into the core material, and the puncture is greater than 1/2 inch, repair as follows and as shown in Figure 10-12 or 10-13.

1. On the panel surface, lay out the area to be removed. A flush patch may be cut first (Step 8) and used as a template to trace cutout area on the panel to insure that the patch will fit snugly in the cutout.

2. Using eye protection, cut out damaged skin using a router chucked with a 1/4 inch drill, a pin router bit, a fine tooth keyhole saw or a circular saw.
3. Cut and remove the core material to the opposite skin using a knife.
4. If both skins are damaged, cut out damage area on opposite side of the panel. Second cutout should be identical to and directly opposite the first cutout.
5. File cutout edges smooth using a fine cut, half round file or rotary file in a 1/4-inch drill.

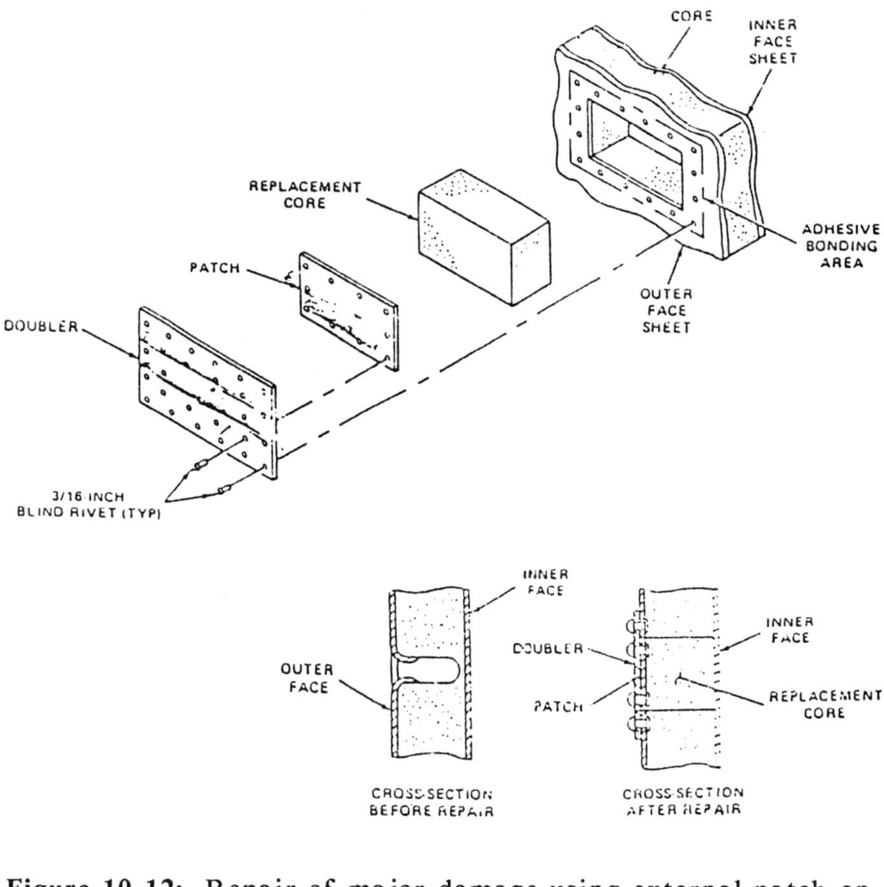

Figure 10-12: Repair of major damage using external patch on one skin only.

6. Remove any paint from the surface around each cutout for a distance of 1-1/2 inches (up to 2 inches for large cutouts). Paint may be removed by sanding.
7. If only one skin has been cut out, clean bonding (interior) surface of the opposite skin by sanding until smooth. NOTE: If cutouts have been made to both sides of the panel, Steps 8 through 24 apply to both skins.
8. Prepare a patch to fit into the cutout in the skin. Patch must be the same thickness and material as the panel skin.
9. Prepare a doubler patch which is one inch larger on all

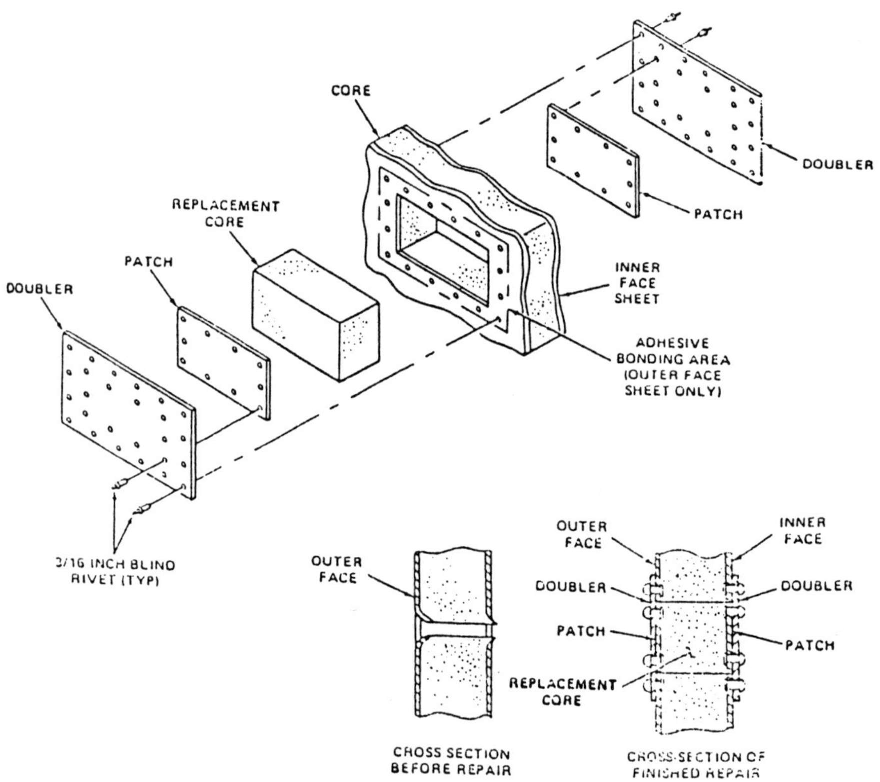

Figure 10-13: Repair of major damage using external patch method for hole in both skins.

sides than the cutout (two inches for large cutouts). The doubler patch should be of the same material as the panel skin and should be 0.032" to 0.064" in thickness.

10. Lay out rivet hole pattern in the patch and doubler. Rivet holes should be 1/4 to 1/2 inch from edges and one inch apart. (Hole spacing may be increased to two inches when the distance between two corner holes exceeds 20 inches.) Drill holes using a #11 or #12 drill bit. Eye protection should be used while drilling.

11. Center patch on doubler and drill rivet holes in doubler using drilled holes in patch as a guide. Secure patch to doubler temporarily with 1/16-inch rivets or sheet fasteners.

12. Center doubler (with patch temporarily fastened) over the cutout in the panel, making sure the patch fits snugly and flush in the cutout.

13. Using the holes in the doubler as a guide, drill holes in the diagonally opposite corners. Then drill the other two corner holes. Cleco clamps may be used to hold the patch in place, or rivets may be temporarily inserted into each hole as it is drill to keep the patch from moving. Proceed to drill the remaining holes. Use a drill and pin as a stop by placing the drill inside the pin and setting the drill in the chuck so that 7/16 to 9/16 inch if the drill is exposed when the pin is against the chuck jaws.

14. Remove doubler and patch. Remove all metal chips and remove burrs from drilled holes.

15. Cut a section of new foam core material to fit snugly in the core cavity. When machining foam core, be careful not to crush foam, since this may result in water absorption. Do not use a planer to machine foam. Core must be flush with inner skin surface when installed. Use a single piece of foam core.

16. Prepare the faying surfaces of the panel, patches and doublers by sanding with 400 grit sand paper until all of the oxide layer has been removed.

17. Wipe the surface with a clean white cotton cloth dampened with safety solvent. Wear clean cotton cloves when handling cleaned parts.
18. Prepare a sufficient quantity of the adhesive, mixing in accordance with the manufacturer's instructions.
19. Apply adhesive liberally to all surfaces of new core and to the cavity and position the core material in the cavity.
20. Apply a thin even layer of repair adhesive to the bonding surfaces of the skin, patch and doubler. Position the patch in place with sheet fasteners. When repairing an external skin, before installing rivets inject sealant into drilled holes to prevent moisture from entering panel.
21. Rivet patch to doubler and doubler to the panel skin using rivets that have been dipped in sealant. (See Figure 10-15 for description of rivets.)
22. Wipe excess adhesive from outer surface of the repair. Do not use solvent to remove adhesive.
23. Allow the adhesive to cure in accordance with the manufacturer's instructions.
24. Coat the edges of the doubler and rivet heads with sealant.
25. Refinish as required.

Method No. 10-5: Repair of Extensive Panel and Core Damage Using Liquid Foam Core Materials (Ref.10-2)

Liquid core material may be used instead of solid core material (as in method No. 10-4) when damage does not involve a structural member. The flush fitting patch may be omitted, since the liquid foam material, when set, will be flush with the external surface of the panel skin. When damage involves a structural member, use Method No. 10-6 or 10-7, as applicable.

Method of Repair:

1. When both the inner and outer skins are damaged, repair the inner skin in accordance with the appropriate method list above before proceeding with the core repair. The panel should be positioned so that the repair surface is horizontal.

WARNING: The use of protective equipment is required when working with liquid foam material.

2. Cut 3/4 inch plywood to approximately the same dimensions as the area of foam to be replaced. Plywood should slightly overlap the cavity.
3. Wrap the plywood in a sheet of plastic release film to prevent the foam material from adhering to the plywood.
4. Cut randomly spaced relief holes two three inches in diameter in the plastic-wrapped plywood.
5. In accordance with the manufacturer's instruction, mix the amount of liquid foam material required to cover the bottom of the cavity to a depth of 1/4 inch. Spread evenly.

NOTE: Expansion of the foam material occurs in 15 to 30 seconds. Before proceeding, make sure that all required materials are available and ready.

6. Immediately cover filled area with plastic-wrapped plywood. Place sandbags or other weights on the plywood to hold it firmly in place. Do not cover relief holes.
7. Allow one hour for foam material to expand and harden. Clean off excess foam from relief holes and carefully remove plywood.
8. Clean surface of foam material and make sure it is level with external surface surrounding cutout.
9. Continue with appropriate method for the replacement of the external patch.

Method No. 10-6: Repair of Panel Damage Spanning an Internal Structural Member (Ref. 10-2)

Panel damage that spans a structural member should be repaired in accordance with Method No. 10-3 or 10-4 as applicable except that the following also applies:

Method of Repair:

1. Liquid foam core material may be used instead of the solid foam core, or the solid core may be replaced as in Method No. 10-4. When repairing an outer skin and using liquid foam material, an aluminum shim the width of the structural member and of the same thickness as the skin shall be used in place of the flush fitting patch (Figure 10-14). The shim shall be installed before applying the liquid foam material. When both the inner and the outer skins are damaged, the inner skin should be repaired in accordance with Method No. 10-4 before proceeding with the core repair. Refer to Method No. 10-5 for repair using the liquid foam core material.
2. Make sure that the structural member is sound before making core and skin repair. If the structural member is damaged, repair it in accordance with Method No. 10-7.
3. If rivnuts have been installed in the structural member in the area requiring repair, drill clearance holes through the patch (or shim) and doubler, as applicable, to allow access to rivnuts and to allow the patch (or shim) to fit flush. (See Figure 10-16 for examples of rivnuts.)
4. Secure patch (or shim) to structural member by use of closed-end blind aluminum rivets (dome head or countersunk). Before installing rivets, inject sealant into the rivet holes and dip rivets in sealant to prevent dirt and moisture from entering the panel. (See Figure 10-15 for description of rivets.)

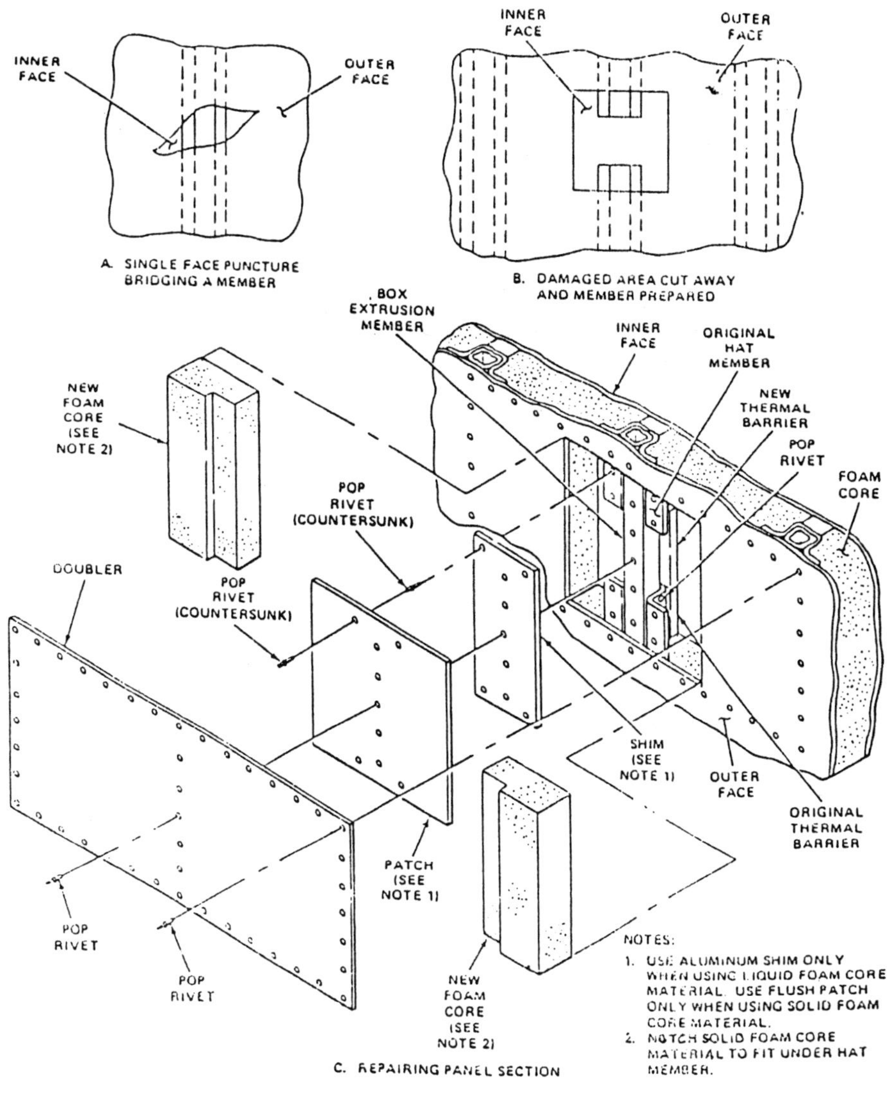

Figure 10-14: Repair of external skin of foam core panel with damaged structural member (Ref. 10-2).

CLOSED-END "POP" RIVETS

DOME HEAD | COUNTERSUNK

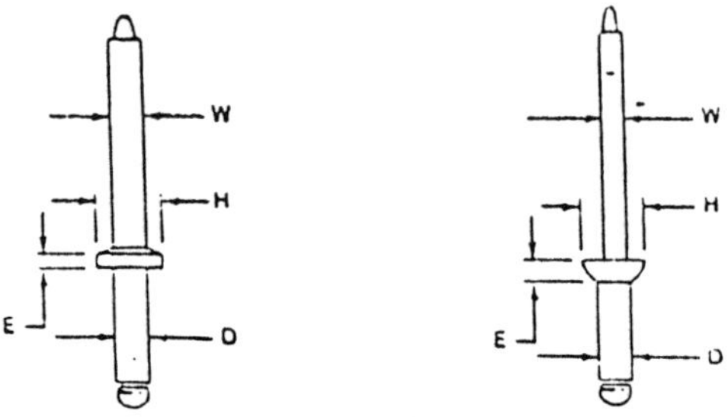

	D	H	DOME HEAD				120° COUNTERSUNK	
			ALUMINUM CLOSED-END WITH STEEL MANDREL		ALUMINUM CLOSED-END WITH ALUMINUM MANDREL		ALUMINUM CLOSED-END WITH STEEL MANDREL	
			E	W	E	W	E	W
NOMINAL RIVET DIAMETER	NOMINAL BODY DIAMETER	NOMINAL HEAD DIAMETER	NOMINAL HEAD WEIGHT	NOMINAL MANDREL DIAMETER	NOMINAL HEAD WEIGHT	NOMINAL MANDREL DIAMETER	REF HEAD WEIGHT	NOMINAL MANDREL DIAMETER
1/8	.125	.236	.036	.064	.036	.072	.035	.084
5/32	.156	.312	.051	.086	.051	.090	.050	.086
3/16	.187	.375	.066	.104	.066	.108	.060	.104
1/4	.250	.500	.084	.144	–	–	–	–

(TYPICAL EXAMPLES)

CLOSED-END RIVETS

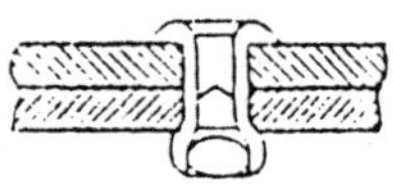

A D 4 4 A H

RIVET BODY MATERIAL
A = ALUMINUM
HEAD STYLE
D = DOME HEAD
K = COUNTERSUNK
RIVET DIAMETER IN 32NDS OF AN INCH
EXAMPLE 4 EQUALS 4/32NDS or 1/8" BODY DIAMETER.
MAX. GRIP IN 16THS OF AN INCH.
EXAMPLE 4 EQUALS 4/16THS OR 1/4" MAXIMUM GRIP RANGE.
MANDREL MATERIAL:
A = ALUMINUM
NO LETTER = CARBON STEEL
CORE DESIGN:
H = HOLLOW CORE
S = SOLID CORE

Figure 10-15: Description of closed-end rivets used in repair (Ref. 10-2).

Method No. 10-7: Repair of Foam Cored Panel with Damaged Structural Member (Ref. 10-2)

A structural member that is sufficiently deformed to cause distortion of the opposite skin will normally be considered unrepairable.

Method of Repair:

1. On the panel skin, lay out the area to be repaired. A flush patch to be used in Step 14 may be cut first and used as a template to trace the cutout area on the skin, ensuring that the patch will fit snugly in place.
2. Cut out the damaged area using a router bit, a fine tooth keyhole saw or a circular skin saw. Eye protection should be used when cutting and/or drilling. Be careful not to cut the structural member while cutting the skin.
3. Cut and remove core material to the opposite skin using a knife.
4. If both skins are damaged, cut out damaged area on opposite skin. The second cutout should be identical to and directly opposite the first.
5. File cutout edges smooth using fine cut, half round file, or rotary file.
6. Remove paint from panel surface around each cutout for a distance of 1-1/2 to 2 inches by sanding with 400 grit sandpaper.
7. If only one face sheet has been cut out, clean the bonding (interior) surface of the opposite skin until smooth.
8. Cut damaged member, and thermal barrier if one is present, so that the ends project four to six inches into cutout.
9. Fabricate a box extrusion from unpainted 6061-T6 aluminum alloy conforming to specification QQ-A-200/8. Box extrusion should be long enough for both ends to extend beyond edges of the cutout. (See Figure 10-14.)

10. Coat extrusion liberally with adhesive and fit it snugly inside cut member. Center extrusion in cutout area.
11. If a thermal barrier was removed in step 8 prepare a piece of thermal barrier from birch wood conforming to specification SC-A-47794-52. The new piece should be the same dimensions as the piece removed.
12. Coat the new thermal barrier with adhesive and fit it snugly in place where the original was removed.
13. Rivet box extrusion to the original structural member. Install rivets through the sides of the original member.

    NOTE: Repair outer skin in accordance with Steps 14 through 32 of this method. Repair the inner skin in accordance with Method No. 10-4, Steps 8 through 25. When damage involves both skins, repair the inner skin before proceeding with the core repair.
14. When using solid foam core material, cut a patch to fit cutout in skin. When using liquid foam core material, cut a shim the same dimensions as the exposed section of the structural member. The patch or the shim must be the same thickness as the panel skin.
15. Cut a doubler which is one to two inches larger on all sides than the cutout.
16. Lay out the rivet hole pattern in the patch (or shim) and the doubler as shown in Figure 10-14. Rivet holes should be 1/4 to 1/2 inch in from edge and one inch apart. Hole spacing may be increased to 2 inches when the distance between two corner holes exceeds 20 inches. Drill holes using a #11 or #12 drill bit (#12 or #30 bits for large patches). Drill two diagonally opposite corner holes first, then the other two corner holes, then drill the remaining holes.
17. Drill rivet holes in panel skin, using holes drilled in the doubler as a guide. Use a drill and pin as a stop by placing drill bit inside the pin and setting bit in the chuck so that 7/16 to 9/16 inch of the drill bit is exposed when the pin is set against the chuck jaws. Temporarilv

insert a 3/16-inch sheet fastener or rivet in each hole as it is drilled to keep the doubler from moving.

18. Remove the doubler. Remove all metal chips and remove the burrs from drilled holes.
19. Prepare the faying surfaces of the skin and the doubler for bonding by sanding with 400 grit sandpaper.
20. When using liquid foam core material, rivet shim to the structural member as shown in Figure 10-14, using countersunk head rivets. (See Figure 10-15 for description of rivets.)

    NOTE: When using liquid foam core material, omit Steps 21, 24, and 25 and repair core in accordance with Method No. 10-5 before proceeding.

21. Cut section of new core material to fit snugly in to core cavity. Core material must be flush with both inner surfaces of the skins when installed. Use a single piece of foam.
22. Wipe all faying surfaces with a clean white cotton cloth dampened with safety solvent. Wear clean cotton gloves when handling clean parts to be bonded.
23. Prepare the repair adhesive in accordance with the manufacturer's instructions.
24. Apply adhesive liberally to all surfaces of the new core material and the cavity and position the core material in the cavity.
25. Rivet the patch to the structural member as shown in Figure 10-14, using countersunk head rivets (see Figure 10-15).
26. Apply a thin even layer of adhesive to the bonding surfaces of the skin and doubler. Position doubler in place with rivets.
27. When repairing an external skin inject sealant into the drilled rivet holes, and dip rivets into sealant before riveting. Rivet the doubler to the skin.
28. Rivet the patch (or shim) and the doubler to the structur-

al member and the box extrusion to reinforce the repaired section.

29. Wipe excess adhesive from outer surfaces of the repair.
30. Cure the adhesive in accordance with the manufacturer's instructions.
31. Coat edges of doubler and rivet heads with sealant.
32. Refinish as required.

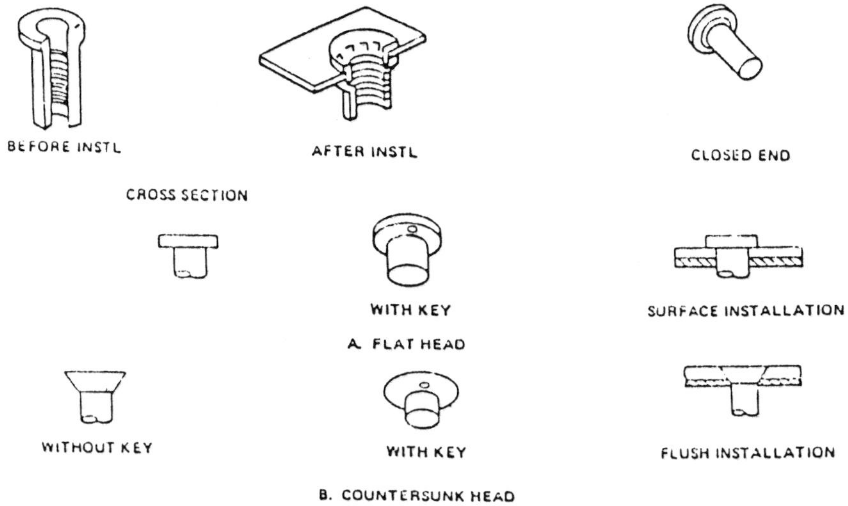

Figure 10-16: Examples of rivnuts (Ref. 10-2).

BALSA CORE SANDWICH PANELS

The repair of balsa wood cored structures can be accomplished by modifying the techniques used for the other sandwich panels. The following methods are modifications of methods previously described.

Method No. 10.8: Repair of a Puncture - Flush Patch Method. This method is a modification of Method No. 10.1.

Method of Repair:

1. Mark a rectangular area that is one (1) inch larger on all sides than the size of the puncture and its surrounding dent or other damage.
2. From this outline determine the size of the rectangular patch panel and cut it to size using the same material and thickness as the original panel material. (See Figure 10-17.)
3. Place the patch panel over the repair area, centrally located within the outline, and scribe its outline on the panel to be repaired.
4. Using a Dotco saw, pin router or other suitable tool, cut through the damaged skin along the inside edges of the scribed outline.
5. Using a sharp knife or other suitable tool, cut through the damaged core following the previously made cuts in the skin.
6. Remove the punctured skin and balsa core material.
7. Using a hand file, trim the edges of the cutout so that they are straight, and so that the cutout is 1/16" larger than the patch panel. (See Figure 10-18.)
8. Using a suitable off-set tool such as a saw blade or under-cutting router bit remove the core and adhesive from the back side of the edges of the cutout. (See Figure 10-18, Section A-A.)
9. Prepare a replacement core insert that is the same size (length and width as the cut out. The balsa core must be dried. The fit should be snug. The material should be of the same thickness and material as the original.
10. Prepare 4 framing strips of the same material and thickness as the ruptured skin. These strips must be 1" wide and the lengths are as follows: 2 ea. 1" longer than the patch panel width, 2 ea. 1" shorter than the patch panel height.
11. Remove core material on the four edges of the replacement insert so that the surface of the framing strips is

even with the face of the core. (See Figure 10-19, Section B-B.)

12. Remove any paint from the cutout edges using a method outlined in Chapter 4 to a width of 1 inch.

13. Prefit the balsa core insert into the cutout area, check for snug fit, remove and set aside.

14. Place one of the longer framing strips at the top edge of the cutout as shown in Figure 10-19. Continue this step using the two shorter strips on the sides and the other long strip at the bottom. Make sure that all the strips are tightly butted against each other.

15. Remove framing strips. Place each part so it can be returned to its original position.

16. Remove all foreign materials such as dust, chips, shavings, etc., from the cutout area and the surrounding panel surface.

17. Using safety solvent, thoroughly clean all surfaces of the

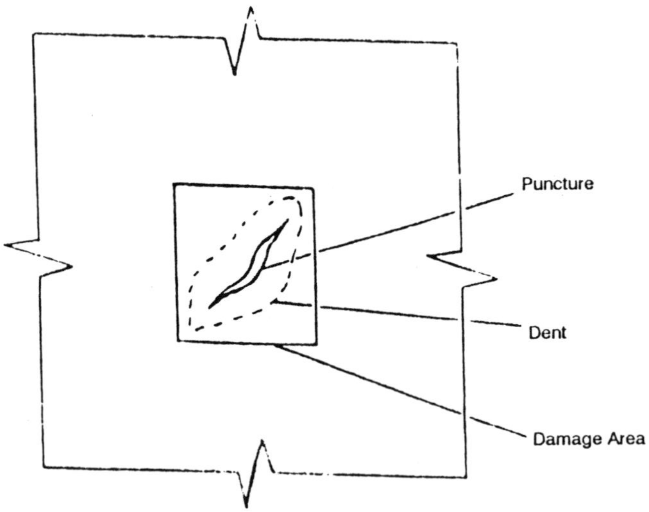

Figure 10-17: Lay-out of damaged area.

patch panel, framing strips and the surrounding panel surface.

18. Mix a repair adhesive (See Table 2-1) in accordance with the manufacturer's instructions.
19. Apply adhesive to all inside surfaces of the cutout area and install the replacement balsa core.
20. Lightly coat the framing strips with adhesive and slide into place around the edges of the cutout.
21. Apply adhesive to the remaining surface of the replacement balsa core and the back side of the patch panel.
22. Install the patch panel.
23. Remove all excess adhesive squeezed out during the installation process and cure the adhesive in accordance with the manufacturer's instructions.
24. Fill the 1/32" gap between the edges of the patch panel and the cutout with adhesive.

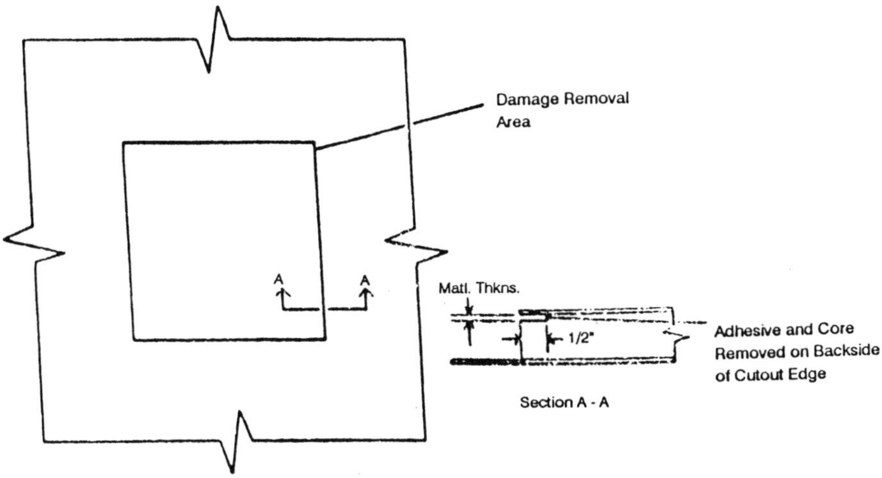

Figure 10-18: Cutout of Damaged Area.

25. Allow the adhesive to cure and the sand smooth using an oscillating sander.
26. Refinish as required.

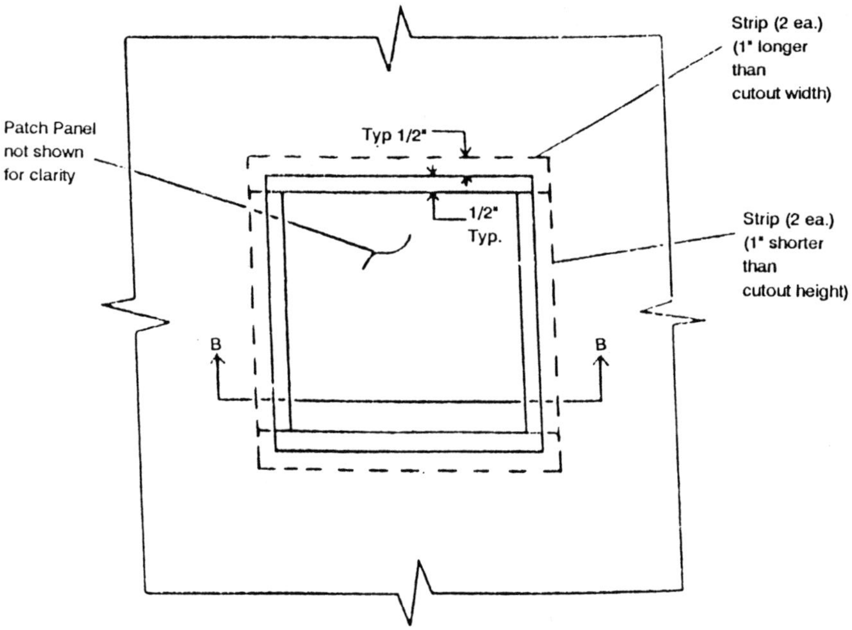

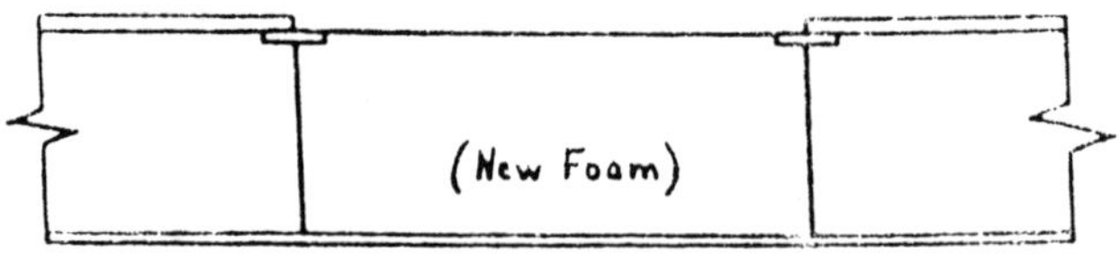

Section B-B

Method No. 10.9: Repair of a Through Hole - Flush Patch Method. This method is a modification of Method No. 10-2.

Method of Repair:

1. Mark a rectangular area that is one (1) inch larger on all sides than the size of the puncture and its surrounding dent or other damage.
2. From this outline determine the size of the rectangular patch panel and cut it to size using the same material and thickness as the original panel material. (See Figure 10-20.)
3. Place the patch panel over the repair area, centrally located within the outline, and scribe its outline on the panel to be repaired.
4. Using a router or other suitable tool, cut through the damaged skins along the inside edges of the scribed outline.
5. Remove the punctured skin and core material.
6. Using a hand file, trim the edges of the cutout so that they are straight, and so that the cutout is 1/16" larger than the patch panel. (See Figure 10-7.)
7. Using a suitable off-set tool such as a saw blade under-cutting router bit remove the core and adhesive from the back side of the edges of the cutout. (See Figure 10-21, Section A-A.)
8. Prepare a replacement balsa core insert that is the same size (length and width) as the cut out. The fit should be snug. The material should be of the same thickness and material as the original. The balsa wood core must be dried.
9. Prepare 4 framing strips of the same material and thickness as the ruptured skin. These strips must be 1" wide and the lengths are as follows: 2 ea. 1" longer than the patch panel width, 2 ea. 1" shorter than the patch panel height.

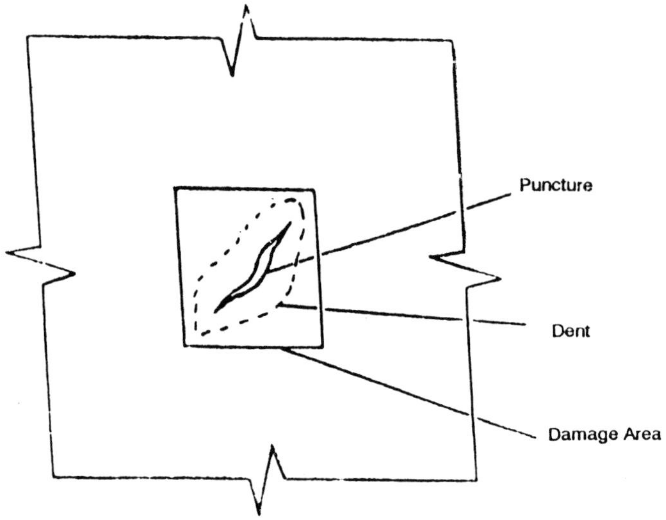

Figure 10-20: Lay-out of damaged area.

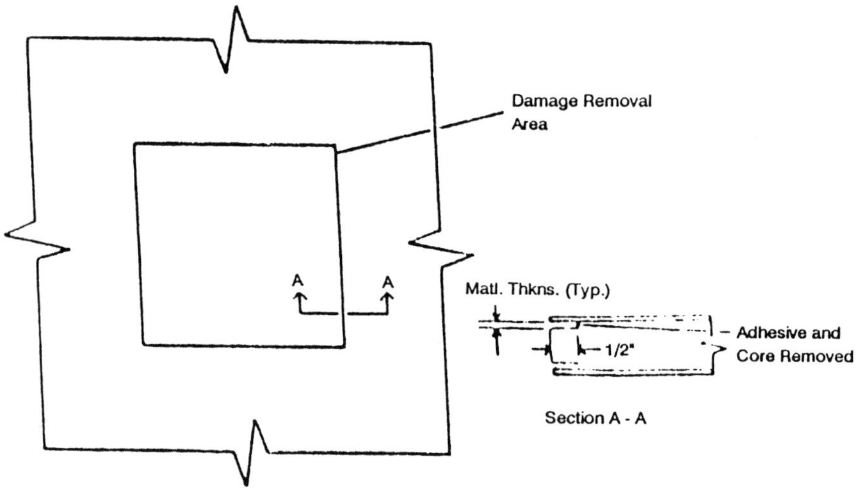

Figure 10-21: Cutout of damaged area.

10. Remove foam on the four edges of the replacement insert so that the surface of the framing strips is even with the face of the core. (See Figure 10-22, Section B-B.)
11. Remove any paint from the edge of cutout edges using a a method described in Chapter 4 to a width of 1 inch.
12. Prefit the core insert into the cutout area, check for snug fit, remove and set aside.
13. Place one of the longer framing strips at the top edge of the cutout as shown in Figure 10-22. Continue this step using the two shorter strips on the sides and the other long strip at the bottom. Make sure that all the strips are tightly butted against each other.
14. Remove framing strips. Mark each part so that it can be returned to its original position.
15. Using safety solvent, thoroughly clean all surfaces of the patch panel, framing strips and the surrounding panel surface.
16. Mix a repair adhesive (See Table 2-1) in accordance with the manufacturer's instructions.
17. Lightly coat the external surface of the framing strips for one side of the cutout with adhesive and place around the edges of the cutout.
18. Install the appropriate patch panel.
19. Remove all excess adhesive squeezed out between the framing strips and the patch panel during the installation process.
20. Apply adhesive to all inside surfaces of the cutout area and install the replacement foam.
21. Lightly coat the remaining framing strips with adhesive and slide into place around the edges of the cutout.
22. Apply adhesive to the remaining surface of the replacement foam core and back side of the patch panel.
23. Install the patch panel.
24. Remove all excess adhesive squeezed out during the

installation process and cure the adhesive in accordance with the manufacturer's instructions.

25. Fill the 1/32" gap between the edges of the patch panel and the cutout with adhesive.
26. Allow the adhesive to cure and the sand smooth using an oscillating sander.
27. Refinish as required.

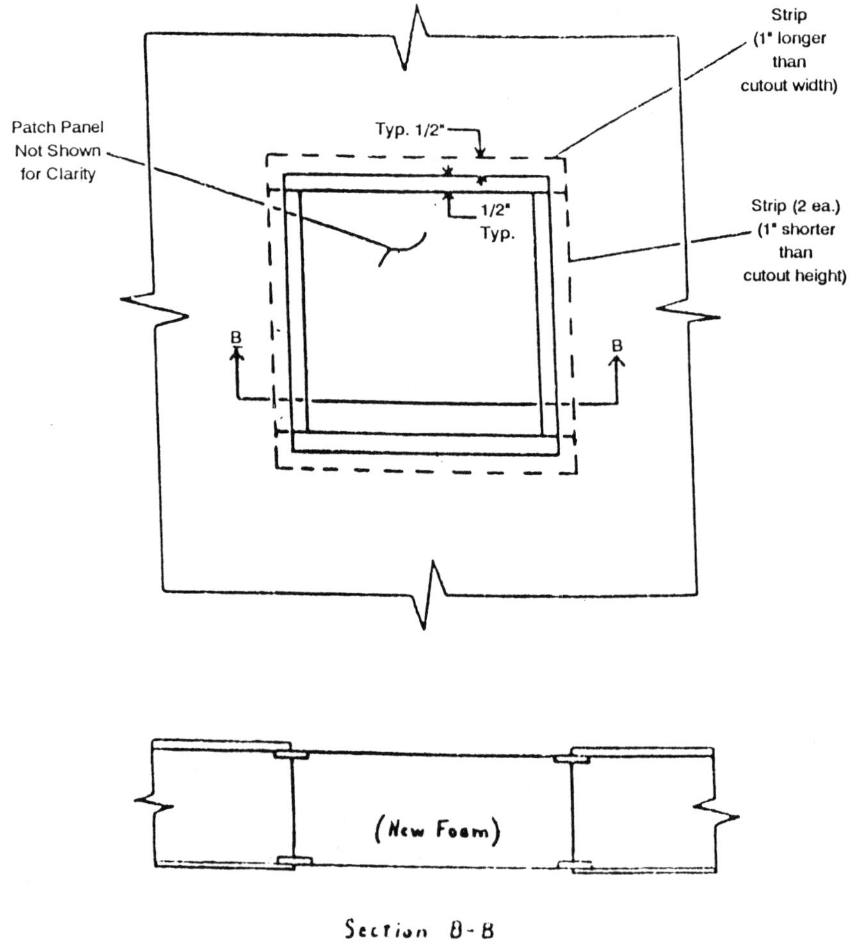

Figure 10-22: Framing strips in repair.

Method No. 10.10: Repair of Composite Faced Structure Containing Major Balsa Wood Core Damage (Hole in Both Skins). This method is a modification of Method No. 6.10

Method of Repair (See figure 10-23):

1. Cover the damage with a solvent resistant pressure sensitive tape.
2. Mask off an area approximately four inches larger on all sides than the area to be repaired.
3. Clean the masked off area with a trichlorethane dampened clean white cotton cloth and remove the paint in accordance with a method given in Chapter 4.
4. Layout the skin and core area to be removed, using a scribe. The layout area should have 1/8 inch minimum radius corners and the scribe lines should not extend past the radii.
5. Drill four 1/4 inch diameter holes in the corners through the panel. (See Table 2-1 for special drill bits).
6. Cut the skin between the holes, on both sides with an appropriate tool.
7. Cut through the core using a sharp knife and remove the skins and the core.
8. Prepare a metal spacer which is the same thickness as the skin and attach it to a larger back plate with a minimum of two rivets. (See Figure 10-24.) Apply a non-silicone release agent to the surface of the back-up plate.
9. Fabricate a replacement balsa core insert from equal material to fit into the removed section.
10. Position the back-up plate into the hole. Seal the outside edges with vacuum bagging putty. (See Figure 10-22, View B.)
11. Mix the repair adhesive (See Table 2-1) in accordance with the manufacturer's instructions.

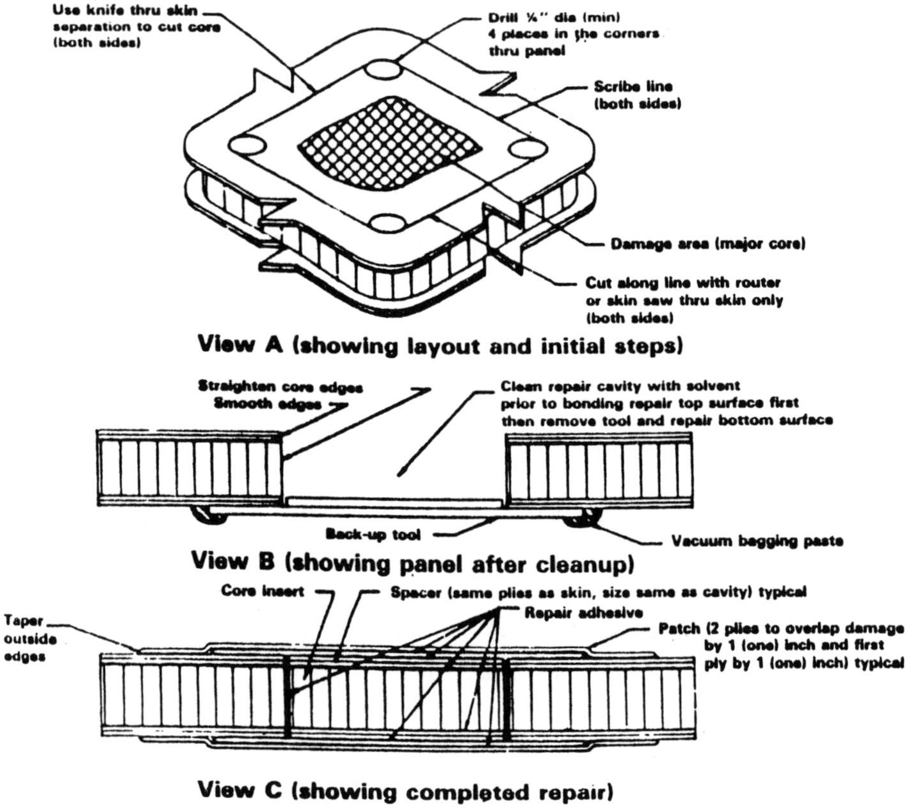

Figure 10-23: Repair of composite-faced structure containing major damage - hole in both skins.

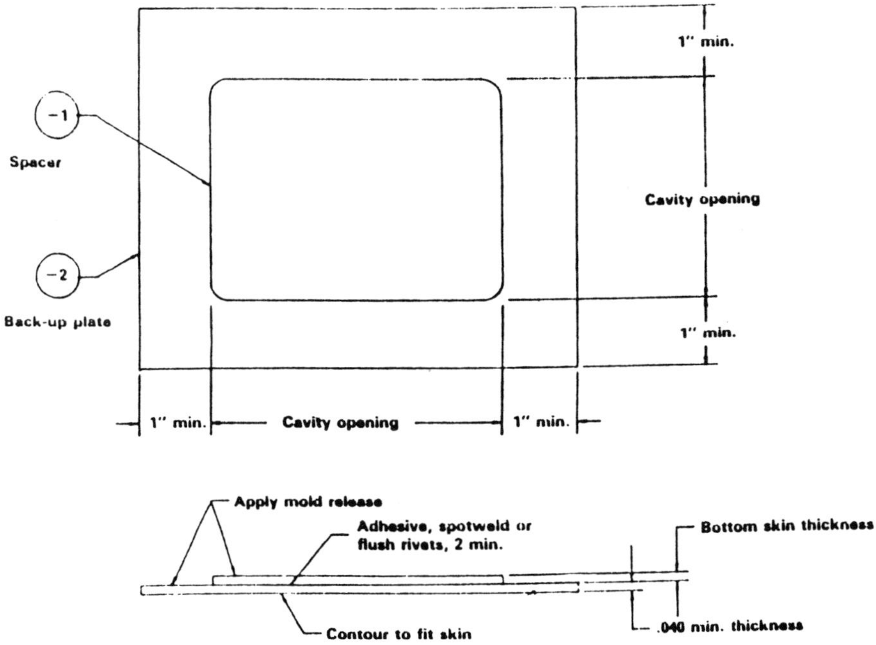

Figure 10-24: Back-up plate for sandwich structure repair.

NOTE: For large Depot repairs a structural film adhesive and paste adhesive may be used.

12. Apply the adhesive to the sides of the cavity and the core insert. Place the insert into the cavity.
13. Cut glass cloth plies the same size as the balsa core insert and of the same number of plies as the original skin.
14. Mix the laminating adhesive (See Table 2-1) in accordance with the manufacturer's instructions.
15. Impregnate the glass cloth plies with the laminating adhesive and position them over the core insert.
16. Add two more plies of glass cloth impregnated with the laminating adhesive. The first ply shall overlap the insert by one inch on all sides and the second ply shall overlap the first by one inch on all sides. (See Figure 10-22, View C.)

17. Cover with a sheet of release film and place an aluminum plate (1/8 inch thick minimum) which is approximately 1/4 inch larger than the last ply.
18. Apply vacuum bag pressure and cure in accordance with the adhesive manufacturer's instructions.
19. Remove the bag and the back-up plate and repeat Steps 14 through 19 for the backside.
20. Sand the repair patch on both sides to obtain smooth and feather the edges.
21. Refinish as required.

REFERENCES

10-1 Gichner Mobile Systems Process Specification Spec No. 3-35 Revision 7, (Dated 4-3-87).

10-2 Department of the Army Technical Bulletin No. 43-0124 "Maintenance and Repair Procedures for Shelter, Electrical Equipment (Field, Intermediate and Depot Levels)", (24 June, 1987).

11

NONDESTRUCTIVE INSPECTION

INTRODUCTION

This chapter contains a description of nondestructive inspection (NDI) methods that are particularly applicable to the quantitative evaluation and detection of defects in adhesive bonded structures. These methods are directed toward the detection of defects which if left undiscovered, might prevent the parts from fulfilling their design functions. Limited discussion will be presented on the qualitative nondestruction of bond strength and durability. The typical or most commonly used techniques are included. Other techniques may be equally applicable. The responsibility for the selection of techniques for a particular application rests with the using facility's Q.C. organization. Further discussions on NDI may be found in SAMPE Handbook No. 1 (Ref. 11-1) and in MIL-HDBK- 337 (Ref. 11-2).

It is important to note that NDI methods are comparative (i.e., they give indications of one area being different from another or the same as another). Reference standards are used to evaluate changes that might indicate a defect. Care must be taken in the interpretation of change indications, since they may be due to changes in the structure. The inspector, therefore must be familiar with the internal construction of the part in order to distinguish between defects and legitimate structural changes.

TERMINOLOGY AND DEFINITIONS

Acoustic Emission (AE) - The class of phenomena whereby transient elastic waves are generated by the rapid release of energy from localized sources within the material, or the transient elastic waves so generated. Acoustic emission is the recommended term for general use. Other terms that have been used in AE literature include (1) stress wave emission, (2) microseismic activity, and (3) emission or acoustic or acoustic emission with other qualifying modifiers.

Acoustic Emission Sensor - A detection device, generally piezoelectic, that transforms the particle motion produced by an elastic wave into an electric signal.

Acoustic Transmission - A technique which combines highly pulsed ultrasonic transducers with an acoustic emission system to detect subtle defects on composites and bonded structures. Also called Acousto-ultrasonics.

Bondline - The layer of adhesive which attaches two adherends. (Also called glue line.)

Couplant (AE) - A material used at the structure-to- sensor interface to improve the transmission of acoustic energy across the interface during acoustic emission monitoring.

Couplant (ultrasonic) - A substance used between the search unit and the test surface to permit or improve transmission of ultrasonic energy.

C-Scan - A means of data presentation which provides a plane view of the material and discontinuities therein.

Dry Contact Acoustic Transmission - Acoustic transmission system that uses a dry couplant.

Defect - A discontinuity, or fault, in a part or material that is detrimental to its serviceability.

Discontinuity - An interruption in the normal physical

structure of a part. It may be in the form of a crack, fold, seam, porosity, disbond, etc. A discontinuity may or may not affect the usefulness of the part.

Delamination - The separation layers in a laminate because of failure, either in the adhesive itself or at the interface between the adhesive and the adherend. Also the separation between layers of a composite laminate which may have been caused by poor resin bonding, impact damage, environmental aging, etc.

Fluoroscopy - Real-time viewing of x-ray radiography.

Holography (Acoustic) - A data presentation system using acoustic waves, analogous to optical holography.

Holography (Optical) - A data presentation system using light waves to form an image.

Immersion Testing - An examination method where the search unit and the material are submerged in water.

Inclusion - Impurities embedded in a material in the forming stage. The inclusions can be deep in the part or near the surface. Normally, they will not have any effect on the strength of the part.

Interface - The boundary between two materials.

Mechanical Impedance - A technique which uses low-frequency mechanical vibrations to locate defects.

Neutron Radiography - A process of making an image of the internal details of an object by the selective attenuation of a neutron beam by the object.

Nondestructive Inspection - Examination of structural parts and components for surface and subsurface defects without damaging the structure.

Pulse-Echo - An inspection methods in which the presence and position of a reflector are indicated by the amplitude and time.

Squirter - A Device using a liquid stream to couple an ultrasonic beam to the test piece.

Through Transmission - A test procedure in which ultrasonic vibrations are emitted by one search unit and received by another at the opposite surface of the material examined.

Transducer - A electro-acoustical device for converting

electrical energy to acoustical energy and visa versa.

Ultrasonic - Pertaining to mechanical vibrations having frequencies greater than approximately 20,000 Hz.

PERSONNEL QUALIFICATIONS

Nondestructive inspection of adhesive bonded structures and composites is relatively new. With the electronic advances of space age technology, new advanced, NDI systems have been developed. The effectiveness of these techniques depends on their proper application by trained and qualified operators. MIL-STD-410D outlines the qualification and certification requirements for NDI personnel. Unfortunately, it does not cover the requirements applicable to adhesive bonded assemblies. The responsibility to ensure that operators are qualified is that of the user. It is recommended that an equivalent of NDI level III personnel determine the technique to be used for pre-and post-damage inspection of composite structures. A level I or II inspector may evaluate the inspection results to determine repair requirements or to accept or reject of the repaired parts.

TEST STANDARDS

There should be at least two basic standards, (1) Composite-to-composite with a known void, and (2) composite-to-core-to-composite with a known void, to check the functioning of inspection instruments. Test standards to calibrate and standardize inspection instruments are essential. An inspection technique is usually as good as the test standards. In ideal conditions, the test standard with known voids of selected sizes should closely duplicate the structure to be inspected. The composite skins or laminates should be of the same materials, thickness and areal weight as the structure to be inspected. Core material should be of the same material, thickness and density. Other variations such as tapered core, steps in the laminate and doublers should be duplicated in the standards. The test standards should fabricate the part being inspected. The defect should be introduced in the same bondline of laminate layer as being inspected in the actual part. When the ideal test standards are not available, some other standards which are similar in structure to the part being tested may be used. However, sandwich standards shall not be used for composite-to-composite or composite laminate structures, or for metal-to-metal structures.

INSPECTION WITHOUT STANDARDS

When standards are not available, a known undamaged area can be used as the standard. To inspect the repaired area compare it to unrepaired area. The instrument readings may differ due to structure changes resulting from the repair. Repeated inspection scans must be conducted using various instrument settings and the inspection should be verified by another type of instrument. A knowledge of the part configuration (i.e., adhesive thickness, composite material, etc.) is essential in the interpretation of the test results.

PRE-REPAIR INSPECTION

Most mechanically damaged areas can be detected visually, however, the extent and the outline of any damage requires a careful evaluation. The following guidelines should be followed when applicable:

1. Outline the damaged area after visual examination.
2. Verify and revise the outline by conducting a tap test.
3. Verify further by using a portable instrument such as the Harmonic Bond Tester or the dry contact acoustic transmission which can be very useful in the determination of impact damage in composite structures.

4. If the damage is in a honeycomb panel, an X-ray examination should be conducted to determine the extent of core damage and/or the presence of moisture in the core.
5. If the damage is in potted or core spliced areas, an X-ray is recommended.
6. If the damage is in a thick composite laminate through-transmission ultrasonic inspection may be recommended.
7. If the damage is in the multiple bondline of a honeycomb structure, through-transmission ultrasonic inspection is recommended. The exposed areas must be sealed off to prevent the ingress of water n to the structure during the test.
8. When degradation of the adhesive bond is suspected dry contact acoustic transmission can be used to evaluate the residual bond strength.

POST REPAIR INSPECTION

Post repair inspection should be conducted to determine that no area has been left unbonded and that no additional damage has resulted during the repair procedures. The selection of the inspection technique to be used should be guided by the availability of test standards. The following are guidelines for post repair inspection that should be used where applicable:

1. Conduct visual inspection of the repaired area for obvious defects. Observe for signs of uncured adhesives. This can be determined by observation of the adhesive flash for lack of solidification or tackiness.
2. Conduct a nondestructive inspection with one or more portable instruments.
3. If the repair area contains adhesive potting or core splice adhesives, a X-ray examination is recommended to ensure that all applicable areas have been filled.
4. If the repair area consists of thick laminates or multiple

bondlines, the use of ultrasonic through transmission should be considered.

5. When standards for inspection are not available, the repair area may be compared to an unrepaired area of the same construction. The inspector should be aware that the instrument readings may vary, however, due to any structural change caused by the repair.
6. Repeat the inspection scan using various instrument settings and verify the inspection results with another type of instrument if available.

GENERAL DESCRIPTION OF NDI METHODS AND EQUIPMENT

The various specific NDI methods may fall within the following generalized methodologies:

1. Visual NDI
2. Optical NDI
3. Ultrasonic NDI
4. Acoustic Emission NDI
5. Acoustic Transmission NDI
6. Radiographic NDI
7. Thermal NDI
8. Mechanical NDI

Visual Nondestructive Inspection

Visual NDI is one of the most important and the simplest form of inspection. It is the observation of the material or product to detect gross imperfections or defects. The visual technique is used to determine that the proper materials are used in the repair as well as for the initial pre-repair and post repair inspections of the structure. It involves the identification of the part and the materials of repair, the observation of the gelation and cure of the matrix materials and the adhesives, the physical measurements for size, shape, and flatness, and the observation of visible defects or delaminations. This method is operator sensitive but of such a significant value that the inspectors should be highly qualified. When used in conjunction with adhesive bonding, visual inspection can often aid in the detection of porous adhesive fillets, lack of fillets, lack of adhesive, edge voids and the lack of (or incompleteness) of the cure.

Optical Nondestructive Inspection

Fiber optic instrumentation may be used for inspecting internal components of a composite or metal skinned hollow structure. Use of these techniques require an opening into the structure through which the fiber optic probe may be inserted.

Ultrasonic Nondestructive Inspection

Ultrasonic nondestructive inspection uses pulsed ultrasound at between 2.25 and 10 MHz. The ultrasonic methods are generally divided into immersion and contact techniques, which may be further subdivided into immersion-through transmission or immersion-reflector plate, and contact-through transmission or contact-pulse echo. These methods may be further refined or automated to produce visual recordings of the defect by using C-scans. The reflector-plate techniques are easier to use than the pulse-echo techniques when attempting to produce C-scan recordings of flat surfaces. Special equipment is required for large panels. Contour followers are needed for contoured parts and special squirters are available for parts which can not be immersed into a water tank. ultrasonic methods are adversely effected by destructive wave interference, which is caused by varying adherend or adherend thicknesses. The destructive wave interference may cause an appearance of voids in a structure where there are none.

Acoustic Emission Nondestructive Inspection

Acoustic emission is a term used to describe elastic stress waves produced in solids as a result of the application of stress. Stress is usually applied to the structure by a mechanical force such as bending, tension, or torsion. Thermal stress can be applied by the use of a hot-air gun. The expansion of the structure caused by the application of heat is often enough to produce the required stress. Stress waves are produced by the rapid release of energy within the material. In fiber-reinforced materials, the acoustic emission is generated by cracking the matrix, debonding of the matrix from the fibers, lamination separation, fiber pullout or breakage of the fibers. In a metallic structure, acoustic emissions may be generated by very weak or failed bonds or by corrosion of the metallic core material. Acoustic emission techniques listen to the different sounds made by a structure under stress. Silence is the indication of a good structure. The more sounds that are detected, the more the structure has degraded. This technique may be used for both pre- and post-repair inspections. Recent advances in acoustic emission have demonstrated that it is possible to check the structural integrity of large aircraft like a Boeing 707 without dismantling the aircraft. This type of check-out was not available through the standard industry inspection methods.

Acoustic Transmission Nondestructive Inspection

Acoustic transmission is also known as acousto-ultrasonics. First developed by NASA to study subtle defects in composites, has since been expanded to many other areas of nondestructive inspection. Acoustic transmission can be used to measure variations in the mechanical properties of the structure. The underlying philosophy behind the use of acoustic transmission to measure these variations lies in its great sensitivity to factors which affect the propagation of the stress waves. The pulser is excited by a well defined, broad band, electrical pulse which is transformed into mechanical displacement of the surface. The mechanical displacement launches stress waves into the structure. As these stress waves travel through the structure they are modified by stochastic processes like those that affect the propagation of acoustic emission signals. The stress waves reaching the receiving sensor have been greatly modified and contain information which can be related to mechanical properties and to defects.

Acoustic transmission is different than acoustic emission in that it uses sophisticated signal processing and multi-parameter analysis. The analysis of the additional waveform characteristics allows much more flexibility in choosing the ones which are the most sensitive to the properties being evaluated. Acoustic transmission has been used to measure the effect of impact on a composite structure, the determination of defects within a composite, to inspect for the inclusion of foreign materials such as separator sheet material in a composite, to inspect honeycomb panels and to determine degradation within the structure. Recently (Ref. 11-1), it was shown that acoustic transmission can be used to determine the strength of an adhesive bonded metallic structure. This work employed a system known as dry contact acoustic transmission. The system used a dry contact rolling probe as shown in Figure 11-1. To determine bond strength the probe was placed on the surface to be tested and the signal received by the the acoustic monitor was recorded as energy. The energy numbers were then plotted as a function of the known strength of the specimens. This type of plot is shown in Figure 11-2. Other surfaces were then tested and the energy values were substituted into the equation for the line, and the strength was calculated.

An instrument known as The ATACS Adhesion Tester (AAT) Model Number 7713 is light weight, compact, on-site usable, easy to use, requires no couplant, and requires minimal inspection skills and training. The AAT can be used to determine bond strength, bond degradation, and bond delamination. The AAT can be seen in Figure 11-3.

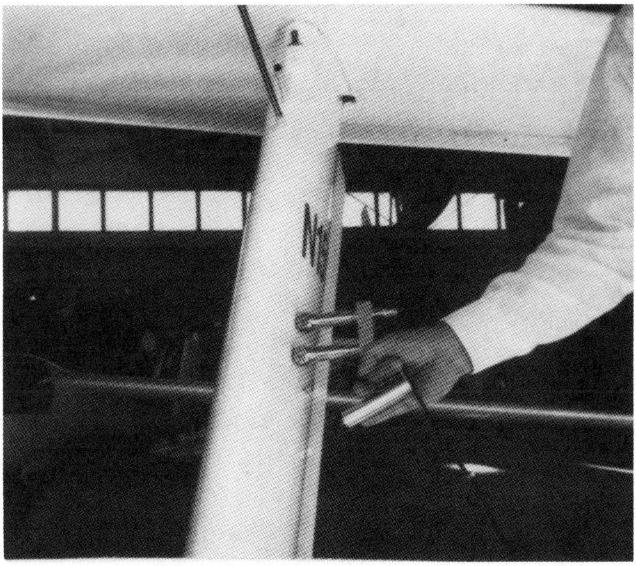

Figure 11-1: The dry contact acoustic transmission rolling probe.

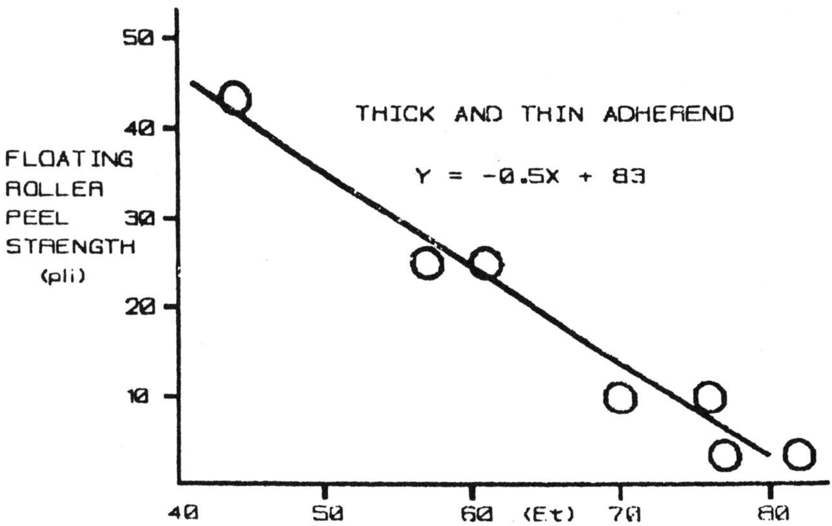

Figure 11-2: Floating roller peel strength vs energy.

Radiographic Nondestructive Inspection

Radiographic nondestructive inspection is the technique of producing a picture on a sensitive surface by a form of radiation other than light. The most widely known form of radiography is the X-ray. Other forms of radiographic inspection include fluoroscopic radiography, neutron radiography, and gamma radiography.

X-Ray Radiography: X-ray radiography is effective for complex geometries that are difficult to inspect by other methods. However, the method must be enhanced by the use of X-ray opaque materials. This method can be used to detect water intrusion into honeycomb structures, cracks, fiber orientation, porosity, inclusions, splice flaws and crushed core.

Fluoroscopic Radiography: Fluoroscopic radiography is real-time viewing of X-ray radiography. This technique is useful for testing bonded panels, composite materials and molded plastic parts. The technique has the same limitations as X-ray radiography, but has the advantage of being in real time so decisions can be made on the spot.

Neutron Radiography: Neutron radiography is particularly useful when components are not X-ray opaque. The hydrogen atoms in water and organic materials are neutron opaque and hence can be imaged to detect water intrusion and defects in composite materials. Metallic components are not neutron opaque, and therefore this technique may be useful for the inspection of adhesive bonds or organic composite structures which may be hidden by a metallic component.

Gamma Radiography: Gamma radiography is similar to X-ray, but at this time is not widely accepted. The gamma rays are obtained from radioactive isotopes such as Cesium 139, Cobalt 60, Iridium 192 and Thalium 70. This technique at present would not be used for repair inspection.

Thermal Nondestructive Inspection

Thermal nondestructive inspection uses the patterns of

surface temperature to detect non-uniformities below the surface. These patterns can be used to determine laminate voids or inclusions, uncured matrix materials, disbonds and cracks. Variations in surface temperature can be displayed by cholesteric liquid crystals. These crystals change color according to their temperature and formulation. Liquid crystals can be formulated to regulate color change for a particular need. Color change can be regulated from -20 to +250 degrees C., with resolution of 0.1 degree C. within a given range.

Large structures such as large nose radomes and some communication satellites are made from composite materials which present some different and unique problems in nondestructive testing. The skins are often very thin and the use of couplants could result in couplant penetration of the laminate surface which may affect the adhesion of the rain-erosion resistant coatings. Photochromic coatings which change color with changes in surface temperature are useful in these applications.

Mechanical Nondestructive Inspection

Tap Test: In the tap test, the surface of the structure is tapped by hand using a tool such as a coin or tapping hammer, such as is shown in Figure 11-4. Voids and disbonds sound hollow in comparison with a void-free area. However, crushed core and an adhesive filled area of core can be mistakenly identified as a good area when tapping inspection is used. This method is very often used as the first inspection method, followed by one or more of the instrumental nondestructive techniques. Care must be exercised in the use of the mechanical tap test as too heavy tapping and too frequent tapping has been known to cause delamination propagation in some metallic structures.

Mechanical Impedance: The mechanical impedance technique is based upon the exciting the surface of the structure with a relatively low-frequency mechanical vibration and the measuring of the response. Excitation may be accomplished by means of a vibrating or tapping pin for testing composite

Figure 11-3: ATACS adhesion tester.

structures. For metallic structures, coil is used which electromagnetically vibrates the metal face sheet of the laminate or honeycomb structure.

The face sheet vibrations create sound waves which are air-coupled to a microphone in the probe. The amplitude of the acoustic response is displayed on the meter. The acoustic amplitude is directly related to the face sheet movement. When the face sheet is well bonded there is little or no movement; over a void the movement is greater. The response from the part being inspected is compared to a previously prepared standard representing acceptable and reject areas. This technique is good for the pre-repair inspection, but for post-repair inspection its use should be verified by other techniques in critical structures.

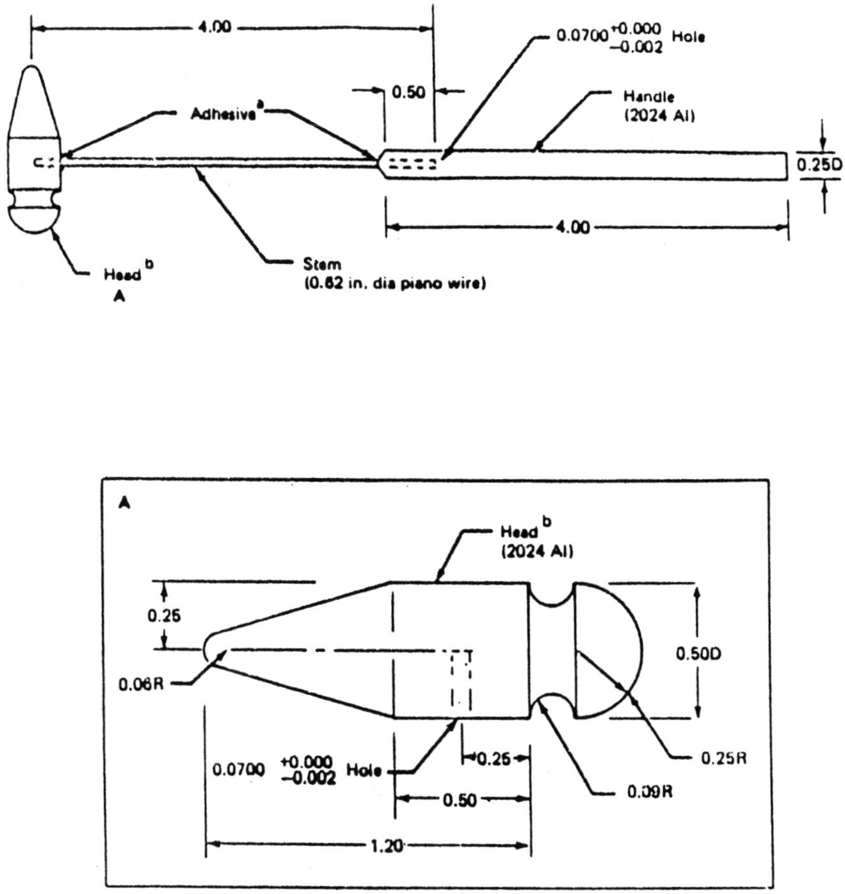

All dimensions in inches

[a] Liquid/paste adhesive may be used if desired. Hole in handle/head may be reduced to provide an interference fit and preclude the need for the adhesive.

[b] 125/All machined surfaces, ref MIL-STD-10

Figure 11-4: Inspection Tap Hammer

Selection of Repair Inspection Methods

Nondestructive inspection methods, both pre- and post-repair, can properly be accomplished by selection of the inspection method using Table 11-1 or 11-2.

Table 11-1
Methods of Nondestructive Testing of Composite Structures

DEFECT >	DELAMINATION	DISBONDS	IMPACT DAMAGE	STRENGTH DEGRADATION	CRACKS	FOREIGN MATTER	CONDITION OF INTERNAL PARTS
ACOUSTIC EMISSION	X	X			X		
ACOUSTIC TRANSMISSION	X	X	X	X	X	X	
RADIOGRAPHY					X	X	X
ULTRASONIC	X	X					X
MECHANICAL IMPEDANCE	X	X					
THERMAL	X	X			X		

Table 11-2
Methods of Nondestructive Testing of Metallic Structures

DEFECT >	DELAMINATION	DISBONDS	BOND STRENGTH	STRENGTH DEGRADATION	CRACKS	FOREIGN MATTER	CONDITION OF INTERNAL PARTS
ACOUSTIC EMISSION	X	X			X		
ACOUSTIC TRANSMISSION	X	X	X	X	X	X	
RADIOGRAPHY					X	X	X
ULTRASONIC	X	X					X
MECHANICAL IMPEDANCE	X	X					
THERMAL	X	X			X		

REFERENCES

11-1 Wegman, R., and Mitchell, J.R., "Evaluation of Adhesive Bond Strength By Nondestructive Testing" 21st International SAMPE Technical Conference, pp. 196-209, September (1989).

11-2 MIL-HDBK-337, Military Standardization Handbook "Adhesive Bonded Aerospace Structure Repair", December (1982).

INDEX